International Series in Operations Research & Management Science

Volume 310

More information about this series at http://www.springer.com/series/6161

Rama Srinivasan Velmurugan • Tarun Dhingra

Asset Maintenance Management in Industry

A Comprehensive Guide to Strategies, Practices and Benchmarking

Springer

Rama Srinivasan Velmurugan
GAIL India Limited
Noida, Uttar Pradesh, India

Tarun Dhingra
School of Business
University of Petroleum and Energy Studies
Dehradun, Uttarakhand, India

ISSN 0884-8289 ISSN 2214-7934 (electronic)
International Series in Operations Research & Management Science
ISBN 978-3-030-74156-3 ISBN 978-3-030-74154-9 (eBook)
https://doi.org/10.1007/978-3-030-74154-9

This Springer imprint is published by the registered company Springer Nature Switzerland AG.
The registered company address is: Gewerbestrasse 11, 6330 Cham, Switzerland

Foreword

I am extremely delighted to write this foreword, not only because Professor Tarun Dhingra has been a mentor, friend, and colleague for more than 15 years. I also believe that the book on asset maintenance with Dr. Rama Srinivasan Velmurugan will be highly useful for students and working professionals who want to have a nuanced understanding of asset maintenance. Asset maintenance has been one of the niche fields which has gained significant attention from the operations management community. Although there exists a rich body of literature on asset maintenance, there is a paucity of good reference books that provide a complete overview of asset maintenance. In response to pressing needs, the authors have invested significant efforts in organizing the chapters that help understand the complex subject with in-depth clarity.

I really enjoyed reading the book chapters of this book. It provides a complete overview of the asset management strategy. I believe this book will become a primer for researchers, teachers, undergraduate students, and working professionals across the globe who are engaged in research and practice of asset management.

Dr. Rameshwar Dubey is a Senior Editor of *International Journal of Physical Distribution and Logistics Management* (IF = 5.212) (ABDC A), an Associate Editor of *International Journal of Information Management* (IF = 5.063) (ABDC A*), Associate Editor of *Journal of Humanitarian Logistics and Supply Chain Management* and Associate Editor of *Benchmarking: An International Journal.*

Liverpool Business School
Liverpool John Moores University
Liverpool, UK

Rameshwar Dubey

Preface

The *Way We Produce* is getting transformed owing to the digitalization of manufacturing, shift in the manner of consumer demand, capturing of data because of automation, Internet, robotics, and optimization of supply chain management. Welcome to evolving Industry 4.0 that has already started changing how we manage our assets in the organizations. Despite such rapid changes worldwide, we are still managing our production assets with thumb rules, norms, and heuristics.

Operations-intensive organizations compete in an industry subject to complex international and national constraints ranging from continuously fluctuating prices to a strict regulatory framework. These organizations need to acclimatize to the ups and downs in the industry always. Therefore, efficient operations across the product value chain are critical in effectively managing their business. The primary support function that can directly impact operations and even stop the operations is asset maintenance.

Despite massive investment in equipment/process/operation, there are a large number of issues in asset maintenance like unplanned shutdown bringing an abrupt stop to the production process, wrong selection of maintenance strategy, non-practice of maintenance strategy, and the cost and impact of equipment safety, upstream supplier issues, and downstream customer issues.

It is experiential that maintenance managers/engineers in the organization do the firefighting even though they follow many asset maintenance strategies. There is a need to systematically approach the maintenance problems and challenges until it arrives at a root cause. A maintenance strategy is primarily concerned with keeping critical systems healthy, fixing the maintenance issue quickly and faster than the last time, identifying the root cause of the repeated problems, and narrowing down to 20% of the maintenance failures that are consuming 80% of business resources.

With the above background, operations-intensive organizations' current business scenario emphasizes treating asset maintenance as a strategic function and developing robust maintenance strategy and building best maintenance practices. This book provides deep insights into asset maintenance strategies, practices, and benchmarking from the research study conducted among maintenance practitioners from operations-intensive organizations.

This book offers comprehensive and updated methodological and practical solutions to operations/maintenance practitioners by answering the following questions:

- How do you select and formulate an asset maintenance strategy?
- How do you build the best maintenance practices at an organization level?
- What is the baseline of operations-intensive organization to existing asset maintenance strategy and practices?
- How to have modern asset management perspectives with a survey questionnaire for assessing a baseline of the organization?

Chapter 1 describes the basic concepts related to asset maintenance management such as maintenance strategy, maintenance practices, maintenance cost, asset classes, asset life cycle management, and sustainable maintenance. This chapter also includes the current maintenance topics and operation research and management science (OR and MS)-related recent developments in asset maintenance. Chapter 2 deliberates on asset maintenance contributing factors and the industry's relevant practical aspects and tacit knowledge built into the theory. Such in-depth knowledge and datasets captured from the maintenance practitioners regarding maintenance strategy and practices followed at an organizational level, significant maintenance challenges, and their impact on operations are presented systematically.

Chapter 3 presents the asset maintenance strategy frameworks and practices available in the practice and application of OR and MS concepts. A detailed section on spare parts management is also provided in this chapter. Chapters 4 and 5 discuss the frameworks for maintenance strategy and maintenance practices, their planning and execution, and intelligent predictive maintenance in Industry 4.0 for operations-intensive organizations in terms of various asset maintenance processes. Based on these frameworks, why the specific maintenance strategy is chosen for a particular equipment/process/operation and how the formulated strategy can be practiced effectively at an organizational level are also explained. A separate section on benchmarking also explains the benchmark study process and framework for benchmarking in the industry.

Chapter 6 presents the modern asset management perspectives with a survey questionnaire for assessing a baseline of the organization in respect of existing asset maintenance strategy and practices and provides a solution map for optimizing the production cost, managing environmental concerns, mitigating safety challenges, implementing Industry 4.0, and achieving maintenance excellence at an organizational level. The development of asset maintenance strategic plan and its implementation in the industry using qualitative and quantitative methods such as case study approach and analytical hierarchy process. Further, implementing operational excellence as a Six Sigma project using the DMAIC approach is discussed in detail in this chapter.

This book is unique in the asset maintenance domain since it provides a comprehensive method for designing suitable asset maintenance strategic plans and developing best maintenance practices for industry practitioners, young maintenance professionals, and engineering/management students aspiring for a profession in the industry. This book includes additional features such as practical scenarios and solutions where maintenance managers must take critical decisions, questionnaires for baseline survey, industry quote boxes, snapshots, and recommendations based on the research study findings.

With combined 45 years of experience of hands-on industry practice and theoretical sound business professorship, we have optimum balance and the right approach for solving the conundrum. This book creates a holistic road map to manage operations-intensive organizations' assets and help you see the transformation required for Industry 4.0.

Noida, Uttar Pradesh, India — Rama Srinivasan Velmurugan
Dehradun, Uttarakhand, India — Tarun Dhingra

Contents

About the Authors

Rama Srinivasan Velmurugan, Bachelor of Engineering with Master of Business Administration (MBA) in Operations Management, has over 28 years of experience in the oil and gas industry. He is a doctorate in oil and gas management from the University of Petroleum and Energy Studies, India. He worked in the technical field, including project management (preconstruction, construction, and commissioning activities of petrochemicals), operation and maintenance of petrochemicals/gas processing plant, and gas pipelines transmission and distribution. He has several publications to his credit in reputed journals such as *International Journal of Operations Management* (Emerald Publishing) and presented papers in Doctoral Colloquium, in international conferences such as the International Gas Research Conference in Denmark and 14th Pipeline Technology Conference in Berlin, and in the areas of asset integrity management, maintenance strategies, and intelligent predictive maintenance and received best paper awards. He is a member of several professional bodies like the Instrument Society of America (ISA), the Indian Society of Training and Development (ISTD), and Chartered Engineer of the Institution of Engineers. Dr. Velmurugan has developed various training modules on asset maintenance, custody flow metering, and city gas distribution maintenance and delivered the sessions in corporate training programs and academic institutions. He has also widely traveled within India and to countries like Canada, Denmark, and Germany to pursue his professional activities.

Tarun Dhingra Dr Dhingra has around 25 years of experience in research, teaching, and industry. He is currently Professor—Strategic Management and Assistant Dean—Research in the School of Business (SoB), UPES, Dehradun, India. Dr. Dhingra completed his doctorate in strategic management from MNNIT, Allahabad, India, as an ICSSR doctoral fellow. He has cleared UGC-NET, June 2005, and completed his MBA (Marketing) from MJP Rohilkhand University Campus, Bareilly, as a science graduate, and published in many reputed academic journals

(WoS, Scopus, ABDC list) of national and international repute. He has won "The best paper award" five times in different reputed platforms such as IIMA, IISC, Bangalore, K J Somiya, Mumbai, SCMS, Cochin, and DMS, IIT Roorkee. His research interests are competitiveness, energy policy, renewable energy, low-cost carrier (LCC), functional strategy, strategic CSR, and stakeholder theory.

List of Abbreviations

AGA	American Gas Association
AHP	Analytical hierarchy process
AI	Artificial intelligence
ALCM	Asset life cycle management
AM	Asset maintenance
ANSI	American National Standards Institute
API	American Petroleum Institute
BCM	Business-centered maintenance
BLE	Bluetooth low energy
BSQ	Baseline survey questionnaires
C and P	Contracts and procurement
CBM	Condition-based maintenance
CMMS	Computerized maintenance management system
CPA	Competitive performance analysis
CPPS	Cyber-physical production systems
CPS	Cyber-physical systems
CSP	Case study protocol
DCS	Distributed control system
DES	Discrete event simulation
DQM	Department Quality Manual
EAMS	Enterprise asset management system
EN	European Standards
ERP	Enterprise resource planning
FMECA	Failure mode effects and criticality analysis
FoF	Factories of future
HMI	Human–machine interface
IaaS	Infrastructure as a service
IAR	Incident analysis report
IAS	Intelligent automation systems

IIoT	Industrial Internet of things
IMOU	Internal MOU
IMS	Integrated management system
IoS	Internet of services
IoT	Internet of things
IPM	Intelligent predictive maintenance
ISO	International Organization for Standardization
KPI	Key performance indicator
LPG	Liquefied petroleum gas
MCDM	Multi-criteria decision making
MCDSS	Multi-criteria decision support system
MOU	Memorandum of understanding
MRO	Maintenance, repair, and operations
MTBF	Mean time between failures
MTTR	Mean time to repair
NG	Natural gas
O and M	Operation and maintenance
OEM	Original equipment manufacturer
OHSAS	Occupational Health and Safety Audit Systems
OISD	Oil Industry Safety Directorate
OR and MS	Operation research and management science
PaaS	Platform as a service
PAS	Publicly available specification
Pd M	Predictive maintenance
PLC	Programmable logic controllers
PM	Preventive maintenance
PMDSS	Plant-level maintenance decision support system
PMS	Performance management system
PNGRB	Petroleum and Natural Gas Regulatory Board
RCA	Root cause analysis
RCM	Reliability-centered maintenance
RFID	Radio frequency identification
ROU	Right of use
RPA	Robotic process automation
RTEM	Real-time equipment monitoring
RTLS	Real-time location system
SaaS	Software as a service
SCADA	Supervisory control and data acquisition system
SOA	Service-oriented architecture
SOAP	Simple Object Access Protocol
SOP	Standard operating procedures
TAR	Trip analysis report

TBM	Time-based maintenance
TPM	Total productive maintenance
UWB	Ultra-wide band
VUCA	Volatile uncertain complex agile
WSDL	Web Service Definition Language
WSI	Web Service Interface

List of Figures

List of Tables

Chapter 1
Asset Maintenance: A Primary Support Function

Abstract This chapter describes the importance of operations in any organization and how the asset maintenance function supports the organization's operational excellence. Asset maintenance concepts cover the major topics such as maintenance strategy, maintenance practices, maintenance cost, asset classes, and asset life cycle management. Modern maintenance management concepts such as e-maintenance and intelligent maintenance aligning with Industry 4.0 are also contemplated in detail. It is imperative from the deliberation that a proactive e-maintenance system using smart prognostic technologies and tools must be part of maintenance strategy implementation to prevent equipment breakdowns.

The major contributing factors in asset maintenance including maintenance tactics, reliability analysis, maintenance performance, planning and scheduling, materials management, organization/human resources (HR), employee empowerment, information technology, maintenance policies/budget, and maintenance challenges are outlined in this chapter. These factors play a crucial role in selecting, developing, and implementing suitable maintenance strategies to achieve best maintenance practices at an organizational level. Further, important research advances or new developments in Operations Research and Management Science (OR and MS) that have affected asset maintenance management in the industry are provided in this chapter.

1.1 Operations: A Lifeline

Operations in the organization comprise all the activities to produce and deliver value to all stakeholders of the organization. The value to the customer is in the form of a product or service. Operational processes convert materials to a product in the industry. Breakthrough changes are happening in the industry; therefore, operations do not fulfill the business requirements in general. The innovative operational process that generates additional value to the customers is the new-age industry's key objective. In the VUCA (Volatility, Uncertainty, Complexity and Ambiguity)

R. S. Velmurugan, T. Dhingra, *Asset Maintenance Management in Industry*,
International Series in Operations Research & Management Science 310,
https://doi.org/10.1007/978-3-030-74154-9_1

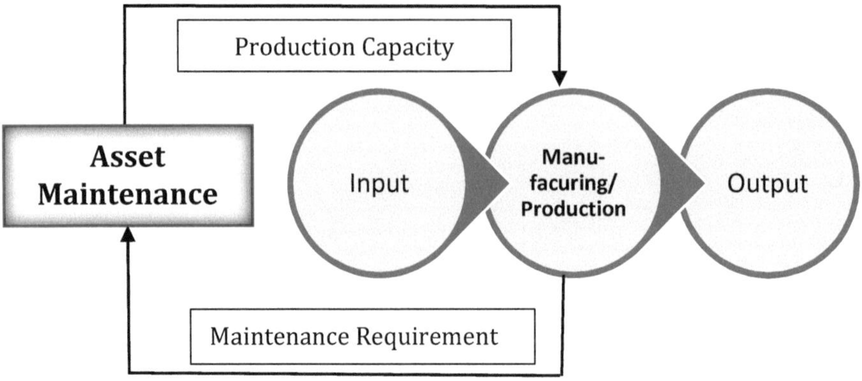

Fig. 1.1 The relation between production and asset maintenance

world environment, the operational processes are changing due to technological advancement in production technology and information technology developments.

Organizations can handle diverse issues due to technology, operational limitations, asset maintenance challenges, strategic and functional planning of the business, and stakeholder involvement. Organizations strive to achieve operational excellence amidst challenges of evolving technologies; industry competitiveness; and safety, health, and environment requirements of the industry by managing the above issues.

Operations must play a significant role in achieving the business profits and growth targets of the organization. The excellent operation management processes can help meet the business requirements and develop a competitive advantage in the industry. Operations also perform a more prominent role in corporate strategy than just enabling the business profits. This role supports the various business processes like strategic planning, quick response in adjusting production targets according to the market conditions, adapting to globalization, and new technologies in advanced operational techniques.

Availability of assets/equipment for operations and ensuring manufacturing efficiency are essential to sustain the production capacity in a continuously operating plant in an organization. Asset maintenance is a support function for operations in the industry to achieve overall quality and business profitability at an organizational level. Manufacturing/production is the main activity that converts the raw material (inputs) into manufactured goods (outputs) with the support of resources such as workforce, materials, and other resources. Asset maintenance is the support activity that helps production function achieve production targets (Gits, 1994). Asset maintenance performs a vital role in the production/manufacturing plant's efficient operations by maintaining the assets/equipment in healthy conditions (Fig. 1.1).

The lack of attention in asset maintenance function in the organizations and the role of maintenance in production lead to substantial issues such as unsafe environment, production loss, and reduction in business profits.

1.2 Role of Asset Maintenance in Operations

The economics of operations broadly hinges on asset maintenance excellence. In normal operations at an organization, maintenance excellence is often overlooked. The operations/production process is considered a primary function. Due to the competitive environment in the industry, asset maintenance is a value generator to the business. Therefore, asset maintenance is acknowledged as an essential function of the business in the industry.

The asset maintenance performance indicators are to be measured thoroughly and monitored for achieving maintenance excellence at an organizational level. Even though such key performance indicators (KPIs) of asset maintenance are being practiced for several years in the industry, a comprehensive and cohesive measurement approach in terms of the organization's current requirement is the major challenge.

> Despite organizations following the good maintenance strategies and practices to achieve business goals, only 7% of organizations are fully convinced with their maintenance performance (Aberdeen, 2006).

In organizations, asset maintenance is a critical support function that enables operational excellence and stimulates organizational growth. The objectives of efficient asset maintenance management are maximizing the production uptime, minimizing asset maintenance cost, quality improvements, zero safety hazards, risk mitigation, and sustainability of operations. These are the tactical requirements of the business in the organization. Optimization of asset maintenance decisions and control over such decisions can be achieved by effectively managing the maintenance challenges.

Asset maintenance managers must make continuous efforts to manage and control asset maintenance decisions efficiently. If they fail in their attempt, the asset/equipment failures lead to substantial financial losses to the organization. Organizations in the agile environment cannot afford such financial losses. Asset maintenance activities are performed in a "firefighting" mode in most organizations. In the absence of a proactive approach, the asset maintenance function can only perform breakdown maintenance after the equipment's tripping. Such a reactive approach resulted in lower business profitability due to lesser asset reliability, asset availability, and productivity.

Random asset maintenance decisions cannot accomplish the optimization of asset maintenance performance. The asset maintenance manager's well-advised and intelligent decisions can only help achieve world-class or benchmarked maintenance performance. All these decisions are to be supported by pertinent datasets in the asset maintenance function. The essentials of asset maintenance excellence are routine maintenance monitoring, identifying improvement areas, providing solutions to the identified improvement areas, and managing the asset maintenance manager's overall leadership.

1.3 Asset Maintenance: Key Concepts

During the past few decades, the asset maintenance function has been acknowledged as a strategic function. Various asset maintenance strategies are available to implement asset maintenance activities efficiently in the organization. The asset maintenance management of assets/equipment involves developing asset/equipment specifications, procurement, installation, commissioning, maintenance, inspection, and replacement.

All these activities of asset maintenance management are to be executed as per the asset maintenance strategy to ensure the best operational performance and profitability of the manufacturing/production plant. The selection of a best-fit asset maintenance strategy is a complex process and crucial in maintenance management. Measuring and quantifying the efficacy of asset maintenance strategy are a significant challenge for asset maintenance managers. Further, asset maintenance management is achieving minimum asset life cycle cost with appropriate process optimization measures. The maintenance decisions impact cost of production, quality of the process, production capacity, and timely delivery of products in line with marketing requirements (Velmurugan, 2016).

An asset maintenance function's primary objective is maintaining the equipment and associated systems in a healthy condition to avoid failures. The equipment performs its function correctly, ensuring a safe environment, and available for uninterrupted plant operations by facilitating quality production, efficient operations, and a sustainable business environment by aligning with corporate strategy.

> The cost associated with maintenance has increased constantly over the decades. The cost of production affected due to maintenance in the range of 15%–70% depends on the type of industry (Ilangkumaran & Kumaran, 2012).

Traditionally, asset maintenance management deals with functional-level issues such as workforce, materials, cost, and other resources. Now, asset maintenance transforms its role in coping with strategic-level problems such as competitiveness, profitability, market requirements, and corporate strategy. One of the challenges of asset maintenance management demands asset maintenance performance improvement continuously. Therefore, asset maintenance is no more a functional activity. It is a strategic activity of the organization to accomplish its corporate objectives to improve its bottom line.

1.3.1 Maintenance Strategy

An asset maintenance strategy is upkeeping the production plants/facilities and equipment in healthy condition. The strategy changes according to the nature of the production process and its operations. It encompasses the failure identification, undertaking equipment repairs based on the nature of the failure, an inspection of equipment parts, replacement with spare parts, and formulating asset/equipment-wise best asset maintenance strategic plan in coordination with all the stakeholders. It explains the relation between occurrences, i.e., breakdown, time delay, and situation, and maintenance action types, i.e., inspection, repair, or replacement. Once maintenance is viewed as a strategic perspective in the organization, there is a need to frame an asset maintenance strategy for performing maintenance functions in the organization. Nonavailability of asset maintenance strategic plan in operations-intensive organizations becomes the barrier for implementing asset maintenance strategies and selecting suitable maintenance tactics like Total Productive Maintenance, Reliability-Centered Maintenance, and Total Quality Management.

Despite the obvious importance of maintenance strategies, a survey among maintenance managers shows that availability of written-down maintenance strategy is 48% and no maintenance strategy exists in 28%. It is also understood that the International Standards Organization (ISO) quality standards are treated as maintenance strategy approaches (Salonen & Deleryd, 2011).

In the industry, asset maintenance professionals are not interested in asset maintenance management's strategic aspects due to their concerns over functional maintenance jobs. There is no interest among the maintenance practitioners in formulating and implementing asset maintenance strategies in the plant/facility at an organizational level since they are overburdened with routine maintenance jobs. Small and medium enterprises are not showing much interest since asset maintenance strategies are more resource-demanding and unwilling to experiment with new asset maintenance tactics such as Reliability-Centered Maintenance and Risk-Based Maintenance. The asset maintenance strategy process consists of asset maintenance strategy formulation, the selection of asset maintenance strategy from the strategies formulated for a specific asset/equipment, and the implementation of the chosen maintenance strategy. All these activities must be carried out comprehensively and strategically.

There are various asset maintenance strategies are availabile to choose from for the organizations in the industry. Adopting such maintenance strategies is crucial since the operations/production process is exceptionally complex due to the automation and mechanization of the assets/equipment deployed in the operations and people dealing with the automated processes. These processes make a considerable impact on managing the quality and capacity of production and asset maintenance costs.

A petrochemical processing plant operates for 345 days in a year with a scheduled annual shutdown maintenance of 20 days. However, industry data shows that production is getting affected for 10–15 days more due to emergency breakdown maintenance in addition to the planned maintenance. Therefore, the plant operates for 330–335 days approx. If it is possible to increase the plant availability by a day, company can profit few million dollars (Velmurugan, 2016).

Despite the best efforts by asset maintenance managers in the organizations, assets/equipment are exposed to failures. Such a scenario in the plant operations keeps asset maintenance managers in a hopeless situation. Therefore, maintenance managers must think differently regarding asset failures and understand the diversity of asset maintenance issues. Merely adopting any asset maintenance strategy does not suffice, and there is a requirement to focus on various aspects of asset maintenance management and contributing factors of asset maintenance.

Operations can be used as a competitive advantage to achieve the industry's business objectives. Maintenance plays a crucial role in operations. Even organizations treat maintenance as an important function nowadays, but it was traditionally considered a cost center. But in today's industrial environment, the strategic perspective of maintenance becomes an essential aspect of its growth.

Despite maintenance managers' consistent efforts in maintaining the assets by applying best maintenance strategies and maintenance practices in organizations, the failures of equipment/process happen, leading to an increase in downtime of plant operations and cost of operations and maintenance. The asset maintenance strategy term is used interchangeably in the research papers, journals, books, etc. with the terms asset maintenance method, asset maintenance policy, and maintenance approach. The asset maintenance methods such as Time-Based Maintenance, Condition-Based Maintenance, Reliability-Centered Maintenance, and Total Productive Maintenance are only part of the asset maintenance strategies. These methods are not maintenance strategies. A strategic asset maintenance plan is a comprehensive process to maintain the asset in a healthy condition during the entire asset life cycle period.

1.3.2 Maintenance Practices

Asset maintenance practices are the best maintenance practices followed by an organization while implementing the chosen maintenance strategy. The mission for best maintenance practices leads to contemporary asset maintenance management that captures technical and operational datasets from the ecosystem and manufacturers' data to effectively implement the selected maintenance strategy considering the asset/equipment's condition during its overall life cycle. A typical contemporary maintenance program attempts to optimize old-style maintenance practices and incorporate condition assessment of the asset, historical information, and other best practices.

Asset maintenance practices are needed to be updated from time to time based on the changes in plant equipment, production technology, and technical expertise required to perform the maintenance activities. The e-maintenance systems (Computerized Maintenance Management System (CMMS), Enterprise Asset Management (EAM), Enterprise Resource Planning (ERP), etc.) must be leveraged in asset maintenance practices to develop asset maintenance strategic plans, resource coordination across the organization, and benchmarked asset maintenance practices infrastructure. Another important aspect of the successful implementation of asset maintenance strategies is optimizing its operational requirements and fulfilling its present business objectives.

1.3.3 Maintenance Cost

The maintenance cost varies based on different maintenance strategies chosen for a specific asset/equipment/process and criticality. The optimization of maintenance costs is the trade-off between the asset maintenance direct expenses and failure prevention cost. The cost prevention measures include regulatory penalties in case of safety and environmental impacts, operational revenue losses due to asset unavailability, and societal implications of asset failures. In safety-critical equipment, the cost of significant accident hazards compensates the cost of industry best practices in preventive and predictive maintenance. For less critical assets/equipment, maintenance costs can become comparable or lower than repair/replacement costs. The initial failure cost of an asset, i.e., immediately after commissioning of the asset, accounts for project cost or a part of operation and maintenance expenses. A detailed discussion on maintenance cost is provided in Chaps. 2 and 4.

One-third of asset maintenance costs are lost due to the uncertainties and inefficiencies implicated in asset maintenance strategic plan (Mobley, 2008).

1.3.4 Asset Classes

Asset classes are generally classified into physical assets and virtual assets. Physical assets are the assets that are physically present in the world. These assets are used in process operations as a physical entity such as gas turbines, compressors, pumps, motors, and reactors. The physical assets can be further divided into two categories:

(1) primary physical assets and (2) secondary physical assets. The primary physical assets are critical equipment used in operations for manufacturing. Secondary physical assets are used for keeping the healthiness of primary physical assets and enable continual operation processes.

Virtual assets are technical data/information used to design, operate, and maintain the physical assets and the information technology solution to collect, store, and manage historical data of physical assets. This information is a valuable resource useful in designing, selecting, and formulating asset maintenance strategies for physical assets. Further, it helps in diagnosing problems in physical assets and improves asset performance. The virtual assets must be comprehensive, accurate, and flexible to maintenance managers/engineers who can access and use the available information. The virtual assets prevail during the entire life cycle stages and are useful in all asset management programs.

Further, virtual assets can be divided into functional, procedural, historical, and commercial categories. The various virtual assets are dealt with the following aspects:

Functional—Design, specifications, and built-in aspects of the physical assets
Procedural—Erection, installation, and commissioning of physical assets
Historical—Physical assets' operation and maintenance throughout the particular asset's life cycle
Commercial—Operation and maintenance costs of physical assets and the financial factors such as warranty, guarantee, and depreciation

1.3.5 Asset Life Cycle Management

Asset life cycle management (ALCM) is a systematic approach in optimizing the physical asset's performance and maximizing its profit in its entire asset life cycle. The asset life cycle study is a must for any maintenance manager/engineer since it describes the procurement stage to the disposal stage processes. The proper maintenance of the assets increases the life and efficiency of an asset. Further, the maintenance manager/engineer needs to decide on the asset's replacement appropriately by repairing over replacement decisions. The various stages of asset life cycle management are as shown in Fig. 1.2.

The new asset's requirement is decided by the operations and maintenance team based on operational requirements in running a plant/facility. In this stage, proper planning considers operational efficiency, maintenance ease, business objectives, and value addition to the process.

After the planning, the next stage is the procurement stage. A proper design of assets, specifications of the asset, and technical requirements must be developed by maintenance managers or consultants/subject matter experts. Further, reviews among the operations and maintenance team members are to be done before placing

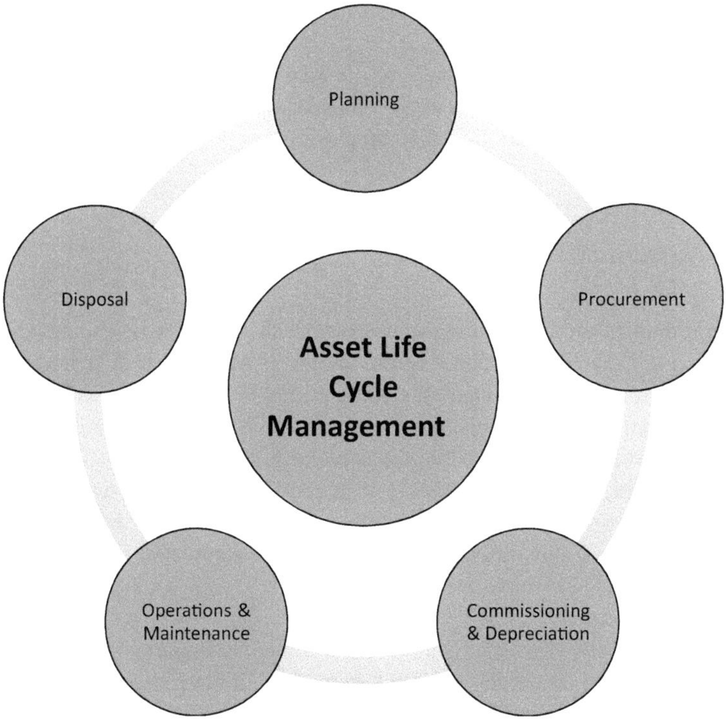

Fig. 1.2 Asset life cycle management stages

a procurement request. After receiving the physical asset at the site, inspection based on the procurement purchase order is a critical aspect of ALCM.

The asset is commissioned after passing the quality inspection. During commissioning, the following aspects are to be kept in mind to ensure smooth operations and fewer failures:

- Installation and commissioning of an equipment/asset must be done as per recommendations by the Original Equipment Manufacturer (OEM).
- Ensure completion of all recommended pre-commissioning activities.
- Commissioning must also follow the industry standards.
- Test run of the asset or equipment is a must.
- Permanent solutions to premature failures after commissioning are an essential aspect of ALCM.

Operations and maintenance is the fourth stage of ALCM, and it is a major time-consuming stage. Therefore, this stage is crucial in deciding the asset's life assessment. All planned maintenance schedules of the particular asset need to be completed as per OEM/licensor recommendations. The missing maintenance activities lead to a significant breakdown and an increase in maintenance costs. Further, balancing between operational objectives and maintenance requirements must focus on improving the asset's life span, which is the asset maintenance manager's primary objective.

The disposal stage is the last stage of the ALCM, which is a critical decision to decide on the non-utilization of an asset. The primary factor should be repairing cost, and this should be more than the replacement cost.

1.3.6 Sustainable Maintenance

The operational environment of organizations is using various assets and equipment for production. The organization's responsibility is to produce or manufacture the products in a responsible manner that suits the environment. Society's sustainable development depends on the sustainable production by the organizations. The asset maintenance function is responsible for achieving the organizational objectives of sustainable production with appropriate asset maintenance processes. The responsibility of asset maintenance managers is to achieve sustainable maintenance. Sustainable maintenance must minimize societal or environmental influence and asset life cycle cost and maximize asset life cycle utility with socioeconomic security (Afrinaldi et al., 2017).

Maintenance activities may lead to environmental impacts in society. A few examples are waste materials due to equipment failure; inefficient fuel/energy consumption in equipment; maintenance wastes like consumables, electronic cards, lubrication oils, and other items; and leakages of gas or liquid due to poor maintenance. The asset maintenance managers must now deliberate on green maintenance practices in the industry. Few such green maintenance initiatives are fuel-efficient equipment, best maintenance job quality delivery, biodegradable lubrication oils, and proper disposal of waste arising out of maintenance activities.

Organizations must focus on energy efficiency in production equipment/assets. The replacement of aging assets at an appropriate time and leveraging the Industry 4.0 technologies for upgrading the sensors for maintaining assets/equipment in healthy condition are the strategies required to achieve energy-efficient asset maintenance systems. Such initiatives are associated with additional costs for the organizations. A trade-off between maintenance cost and sustainable maintenance is crucial for developing an asset maintenance strategic plan and adopting the best maintenance strategies. Asset maintenance practitioners' responsibility is to ensure green and sustainable maintenance activities for a better living society.

1.4 Modern Maintenance Management

The development of modern maintenance management systems emerges from the technological developments of communications systems and information technology (IT) systems. Modern maintenance management systems using e-maintenance applications are Computerized Maintenance Management System (CMMS), Enterprise Asset Management System (EAMS) and intelligent automation systems such as the Internet of Things (IoT) and Cyber-Physical Systems (CPS). The smart predictive technologies and tools developed based on these systems facilitate implementing the asset maintenance strategies effectively. Such proactive, modern maintenance systems maximize asset/equipment availability, reliability, and safety to prevent equipment/asset breakdown and improve product quality to ensure customer satisfaction. E-procurement implementation ensures the timely availability of materials while performing asset maintenance. Further, these tools help carry out continuous asset maintenance assessments and predict asset/equipment behavior. Modern maintenance management techniques with appropriate asset maintenance logistics can prevent equipment or plant shutdowns by improving asset performance.

1.4.1 E-Maintenance

E-maintenance becomes an essential scientific discipline of asset maintenance management. The development of e-maintenance is based on various industrial and academic actors related to technology developments and organizational and management issues of asset maintenance. E-maintenance tools help gather asset maintenance datasets from plant/facility and analyze these datasets using computerized applications that save enormous cost and time due to manual collection of datasets by asset maintenance engineers/technicians. The computer applications used in e-maintenance systems (CMMS and EAMS) are designed to collect datasets of asset maintenance and record all the data about specific equipment/asset in the system for retrieval from the historical database as and when required. Organizations are using the computer applications as mentioned above for asset maintenance data collection and analysis.

There are few challenges in implementing e-maintenance concepts such as skill training on computer applications to the asset maintenance team, replacement of conventional sensors/devices with smart types of sensors, network connectivity, and bandwidth availability. The essential requirements of e-maintenance implementation are the flow of data from the factory level to the computerized system level for recording and performing a study, information about the manufacturing/production logistic operations, and cost incurred due to asset maintenance logistic inadequacies of the production system. All this information is used seamlessly to establish an e-maintenance system for asset maintenance management at an organizational level.

1.4.2 Intelligent Maintenance Systems

Since then, the above-described CMMS has matured around the 1980s and has been implemented in various manufacturing and asset-intensive organizations. However, new solutions available include technological developments in the past decades: Internet of Things, Industry 4.0, and Maintenance 4.0 enable upgradation of CMMS. The CMMS applications are also optimized for mobile devices. Therefore, the key factors influencing modern maintenance systems are Intelligent Predictive Maintenance, dataset availability, and Industrial Internet of Things in Industry 4.0. The maintenance practitioners must equip with suitable intelligent systems to ease their daily maintenance tasks and decide on appropriate maintenance interventions when considering a specific asset/process/plant with overall equipment efficiency.

It is understood that the maintenance practitioners (*refer to box*) are not prioritizing the maintenance jobs and perform the jobs on a day-to-day basis. The missed backlog maintenance jobs are not completed timely since unexpected reactive/breakdown maintenance jobs accumulate and overburden the maintenance team. The entire team gets into this vicious cycle and firefight in performing maintenance jobs. With a smart automation system of maintenance work orders and appropriate interventions cum feedback systems, the maintenance team can be empowered with the new technological tools to focus on what is essential to ensure the asset/equipment/plant's healthiness. Therefore, Intelligent Predictive Maintenance tools equip maintenance practitioners to perform better maintenance job management.

Thirty-five percent of maintenance managers/engineers are aware about the ratio between reactive maintenance and proactive maintenance jobs, and 25% of maintenance managers/engineers are not much concerned about backlog of preventive maintenance schedules (Velmurugan, 2016).

Various datasets are available in the plant environment from automation systems that interface between equipment and human-machine interface (HMI). Few automation systems are Programmable Logic Controllers (PLC), Supervisory Control and Data Acquisition (SCADA) systems, and Distributed Control System (DCS). A modern management computer application software must be integrated with all these automation systems to acquire and utilize the data for taking maintenance decisions on selecting maintenance strategy and adopt suitable maintenance practices. Industrial Internet of Things (IIoT) is a game-changer for the shop floor maintenance team. The maintenance work order generation becomes automated without interventions of the operations team using such technology. The next level of technological improvement is advance predictive maintenance analytics using machine language in the broad framework of Artificial Intelligence. A framework and recommendations regarding Intelligent Predictive Maintenance are discussed in Chap. 5.

Modern maintenance management systems must deal with the following critical aspects while selecting suitable intelligent maintenance systems:

- Modern maintenance management systems' decisions are based on selecting the right datasets to improve performance services. The application of such datasets helps to choose appropriate technology and optimum automation solutions.

- Integration across various plant sites and work centers of the organization across the country is an important aspect of leveraging technology and maximizing modern maintenance management benefits.
- The modern maintenance management system must have a proactive and data-driven approach in designing intelligent maintenance systems. This kind of approach leads to a 35% reduction in the equipment/asset/plant's downtime.

Preventive Maintenance cost – A Snapshot

Preventive Maintenance (PM) Cost/Year = $ 21000
$ 3500 per scheduled maintenance X 6 (frequency is on bi-monthly basis)

PM (using Intelligent Predictive Maintenance) Cost/Year = $ 10500
$ 3500 per scheduled maintenance X 3 (frequency is every 4 months)

1.5 OR and MS Concepts Related to Asset Maintenance Management

This section provides important research advances or new developments in Operations Research and Management Science that have affected asset maintenance management in the industry. The application and impact of these concepts in the industry are discussed in Chap. 3 (Sects. 3.2.2 and 3.5).

1.5.1 MRO Supply Chain Management

Historically, maintenance, repair, and operations (MRO) supply chain management is not given much importance. For various reasons, including MRO supply chain, cost is seen as unimportant by asset maintenance managers/engineers compared to maintenance capital and revenue costs. Asset maintenance managers within the maintenance department manage MRO supply chain jobs. They do not have specific supply chain management experience due to silo organizations and improper decision-making in asset life cycle management.

The low MRO supply chain development affects various aspects of asset maintenance such as process, organization, technology, and inventory systems and the entire maintenance value chain. This condition leads to visible impacts in the asset maintenance management like increased plant shutdown due to nonavailability of spare parts and low store productivity. Further, control and manual interventions increase the error and affect the automated process flow of asset maintenance management system, unnecessary increase in supply chain management cost, and no single point of accountability for MRO supply chain activities.

Maintenance, repair, and operations supply chain management is dealt with by two key departments in the organization like contracts and procurement department and maintenance department. The critical issue of the MRO supply chain is keeping a better inventory control mechanism. Overprocurement of spare parts of equipment adds higher cost to the company, these spare parts may be lying at stores without use, which increases inventory carrying cost. If the procurement department puts restrictions on maintenance engineers, the nonavailability of spare parts may lead to the stoppage of equipment and plant operations. Therefore, the aim is to keep an optimum inventory of spare parts with proper MRO supply chain management. To achieve this aim, organizations implement the Internet of Things (IoT)-enabled spare inventory control using the just-in-time spare parts model. The industry reports show that such an application can result in significant inventory cost reduction, i.e., 30%–35% (Velmurugan, 2016). ISO 8000 standard enables for a collaborative approach among the manufacturers, suppliers, and organizations.

MRO supply chain management requires robust materials data management. Therefore, organizations develop digital strategies for supply chain management optimization using Industry 4.0 technology. The new technology enables efficient business processes by improving spare part traceability with less time and spare parts data management with easy access and reliability. The key factors are contract management, supply agreements, spare parts specifications and demand requirements, and inventory control. The few inventory control aspects are economic order quantity (EOQ), inventory levels, duplicate spare parts, obsolete spares, and procurement lead time. The new technology helps develop predictive MRO supply chain models to optimize the spare parts management business processes.

The effective management of the MRO supply chain can be accomplished by setting the basics right, i.e., identifying the problem areas and planning for improvement in those areas. Once the basics are set right, cloud-based software as a service (SaaS) can be used to manage master data of the MRO supply chain and establish adequate controls of inventory. The automated tools can be used to avoid manual interventions in the MRO supply management cycle leveraging big data platforms. Collaboration with material suppliers, centralized inventory management across the organization and vendor-managed inventory plan further improve the MRO supply chain management.

1.5.2 Decision-Making on Asset Maintenance Strategies

Asset maintenance involves decision-making in the selection of appropriate asset maintenance strategies for equipment. Using the recent developments in Operations Research, the computer-based Decision Support Systems (DSS) can solve the said decision-making problem. The application of DSS models improves the decision-making processes in complex systems to reduce uncertainties and human errors. From simple to Artificial Intelligence, various models are available, using different algorithms and statistical tools like Bayesian. One of the methods is discrete event simulation (DES) for developing models to choose maintenance strategy by understanding intricate interactions between asset maintenance strategies and their impact on asset maintenance implementation.

To select asset maintenance strategies, mix for equipment, an integer linear programming approach is used in the organizations for better resource allocation among the asset maintenance activities for optimum utilization of the available resources. The linear programming approach can also be embedded with Reliability-Centered Maintenance (RCM) to effectively implement the asset maintenance strategies in the plant/facility. Another method is combining the Multicriteria Decision Support System (MCDS) with risk analysis for complex and uncertain asset maintenance systems. This approach helps develop a decision-making framework for handling physical assets in the plant/facility by fulfilling business objectives.

In general, Operations Research/mathematical models (Alsyouf, 2009) used in the industry are failure mode effects and criticality analysis (FMECA), decision trees, multicriteria decision-making (MCDM), and risk analysis. Organizations aim to utilize mathematically based models for selecting the most appropriate asset maintenance strategic plan with both operations and maintenance departments' involvement. One such useful model is the Analytic Hierarchy Process (AHP). This method prioritizes the criteria for asset maintenance strategy selection and ranking among the selected asset maintenance strategies. AHP can be combined with approaches such as quality function deployment (QFD) and benefit of doubt (BoD). QFD helps to align the organizational business objectives with asset maintenance strategic objectives. Further, BoD ensures the robustness of the selected asset maintenance strategy through sensitivity analysis (SA). A detailed process of selecting asset maintenance strategies among the selected maintenance strategies using AHP is described in Chap. 6.

The mathematical modeling is used to identify the assets in terms of single-unit systems and multiunit systems. Further, subclassification is based on the state space of the deterioration process modeled as discrete and continuous condition monitoring, inspection, replacement, repair, and the various types of dependencies that may exist between units within systems. Such classifications are used to optimize asset maintenance policies. Single-unit systems are the systems that are considered an asset or equipment as a single unit for maintenance modeling using various state spaces of the equipment's deterioration process, while in multiunit systems, the individual units or components of an asset/equipment system are modeled separately (Bram & Philip, 2020). The reasons for limited applications of modeling of single-unit and multiunit systems in the industry are as given below:

- Availability of historical data of an equipment/asset over the asset life cycle period is a challenge since the organizations are maintaining limited datasets of an asset.
- Many organizations are using traditional sensors or devices as field instrumentation in production plants/facilities, and during the last few years, smart sensors or devices are replacing conventional sensors. Therefore, data traceability for the entire plant's life is not available with most organizations (Choi et al., 2018).
- Asset maintenance mathematical modeling must consider the production planning and related conditions of the processes. However, such alignment of operations with maintenance for single-unit and multiunit systems plays a significant role in monitoring modeling.

1.5.3 Inventory and Asset Tracking

Inventory management plays a vital role in asset maintenance management to manage the spare parts availability as and when required and retrieval of the spare parts from the stores across the various plant/facility locations of the organizations. Centralized spare procurement helps optimize the spare availability among the stores in the organization's plant locations. Traditionally, materials management in the industry deals with bar code and data collection through maintenance management systems or enterprise management systems. The recent developments like Real-Time Location System (RTLS) in alliance with the Warehouse Management System (WMS) help quickly identify the required spare parts available within the organization and further even determine the availability of spares with the vendor warehouse. The RTLS system uses technologies such as radio-frequency identification (RFID), Bluetooth Low Energy (BLE) beacons, and ultra-wideband (UWB) or Wi-Fi technologies.

RTLS is an indoor positioning system within a plant/factory that tracks assets, spare parts, and production raw materials. Such application enables the organization to reduce plant shutdowns, identify improvement areas in asset maintenance management, and ensure the hazardous plant's safety features. RTLS uses a blend of hardware and software to develop an indoor positioning network within the organization or plant premises. The RFID tags and receivers are placed all over the plant premises; accordingly, the receivers receive the signal's RFID tags. These tags are embedded in equipment, spare materials, tools, and other asset maintenance systems. The transmission signal includes information such as type of the asset/material, location, unique identification number, and status. The receivers are connected to the application software through a hardware system, i.e., a server for centralized processing of the information and storage and retrieval as and when required.

RTLS can be divided into precision-based RTLS and proximity-based RTLS. Precision-based RTLS uses either UWB or Wi-Fi technologies. This system can be well integrated into inventory management and track an asset/equipment's exact location across geographical locations. The cost of technology infrastructure is higher than the proximity-based RTLS. The proximity-based RTLS uses Bluetooth technology. This system can be used within the premises only with a specified distance. The applications of RTLS are the identification of the location of assets/ equipment in a warehouse or plant, logging of equipment records, and retrieval of documents as and when required with accurate traceability of assets and spare parts in the stores. Several Real-Time Equipment Monitoring (RTEM) systems are available to monitor equipment and the development of asset maintenance management models based on the accumulation of historical data in the system using the above technologies.

1.6 Asset Maintenance Contributing Factors

The major contributing factors in asset maintenance are maintenance tactics, reliability analysis, maintenance performance, planning and scheduling, materials management, organization/human resources, employee empowerment, information technology, maintenance policies/budget, and maintenance challenges (Velmurugan, 2016). These factors are depicted in Fig. 1.3.

Maintenance tactics (MT) methods are preventive maintenance/Time-Based Maintenance, predictive maintenance/Condition-Based Maintenance, Reliability-Centered Maintenance, and Total Productive Maintenance. These maintenance methods are generally followed in various industries such as oil and gas, coal, power, automobile, and other manufacturing facilities. The maintenance tactics

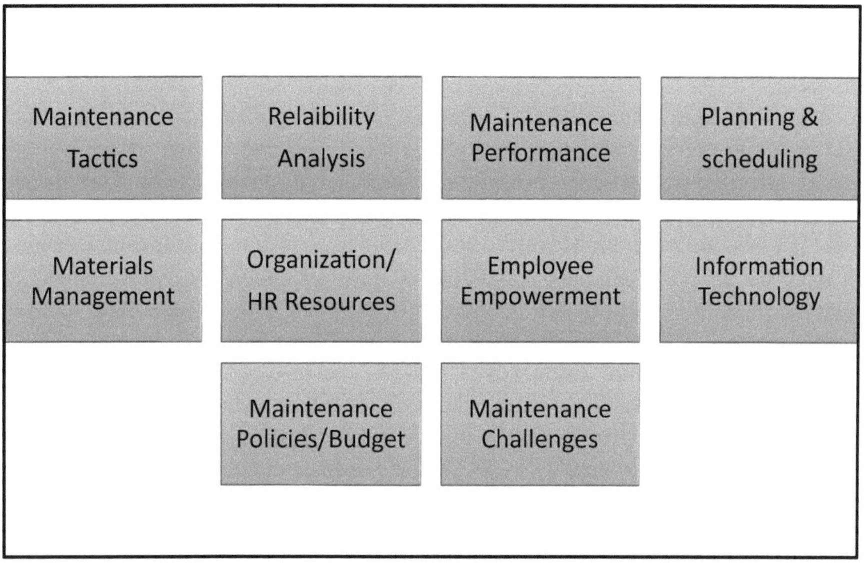

Fig. 1.3 Asset maintenance contributing factors

play a crucial role in designing a maintenance strategy for an organization. Further, the asset maintenance tactic implementation is decided on the criticality of the process and maintenance complexity.

Reliability analysis (RA) plays a critical part in the formulation, selection, and implementation of asset maintenance strategies. Reliability analysis as an overall function that deals with prediction, analysis, prevention, and mitigation of failures over the asset life cycle. It is a scientific process of failure study, identifies the root cause of the failure, and avoids the repetition of such failures. In addition to this, RA also deals with the availability performance of an asset. The availability of an asset/equipment is the time between two consecutive failures of the asset, e.g., the time to restore the equipment's operations. This analysis helps prevent potential failures, working upon design aspects, materials, and maintenance.

Maintenance performance (MP) measurement supports choosing a suitable maintenance strategy for an operation process/equipment. Formulation and selection of maintenance strategies are improved and modified based on the performance measures in maintenance. Organizations need to set benchmarks/targets to evaluate the efficacy of asset maintenance performance and forecast future accomplishment. Such kind of analysis is beneficial to asset maintenance management. One way of implementing this practice is by using maintenance metrics to evaluate the equipment's performance. These metrics should be carefully chosen to achieve business objectives, operational excellence, and maintenance excellence.

Planning and scheduling (PS) is the significant actions of any asset maintenance strategy design and implementation in an organization. The key objective of effective planning and maintenance schedule is to enable the maintenance crew to prepare and perform the maintenance tasks in safe and cost-effective methods without delay in execution. Planning and scheduling deals with a proactive job order management process, job estimation and prioritization, job backlog management, job plan development, scheduling, coordination at the execution level, and tracking the job progress.

Materials management (MM) is an integral part of maintenance strategy formulation and implementation in an organization. In general, on the shop floor, the gap in effective maintenance execution lacks materials management. That means ensuring spares and material availability in case of the requirement are the missing links in asset maintenance. Therefore, it is crucial to consider the following elements in materials management: nonavailability of material, emergency procurement of material, inventory control, and integration of materials management with maintenance planning, which affect the implementation of maintenance strategies.

Organization/human resources (O/HR) is the resource outsourcing strategy that provides flexibility in the maintenance organization so that there is no need to create fixed assets/infrastructure. HR function plays an inclusive role in the formulation of maintenance strategy, selection and implementation of a maintenance strategy, and helping in an organization's maintenance decisions.

Employee empowerment (EE) is a useful management tool in making quick decisions in emergent situations during maintenance. The empowerment of the employees leads to better implementation of asset maintenance strategies at an organizational level. Employee empowerment majorly deals with delegation of power to decide when required, enable a working-level workforce to take preventive and quick decisions in case of emergencies and faster communication to concerned authorities.

Information technology (IT) is a crucial contributing factor in asset maintenance management. IT systems in the plant environment are smart field instruments/ sensors, networking of control systems like Programmable Logic Controllers and Distributed Control System, control systems, and asset maintenance management systems like CMMS and EAMS. IT infrastructure's major role is to bring together all asset maintenance datasets and use them for analysis. The result emerged from asset maintenance data analysis to take appropriate maintenance decisions and optimize the asset maintenance performance. IT systems facilitate digital communication among the asset maintenance managers, operations managers, Original Equipment Manufacturers, and subject matter experts for sharing asset maintenance-related information, experience, knowledge, and skills. Such communication enables problem-solving even though large distances may separate team members. Organizations make investments in IT systems expecting better maintenance practices and their impact on IT utilization.

Maintenance policies/budget (MP/B) strengthens the process during asset maintenance strategy formulation and provides guidelines while implementing the chosen maintenance strategy. This contributing factor performs a vital role during the complete asset maintenance strategy cycle from formulation to implementation at an organizational level. Since the organizations execute asset maintenance to maximize the Return on Investment, determining an appropriate maintenance policy is crucial for sustaining an efficient maintenance environment. The maintenance policy should take care of all the assets and the associated systems deployed in the plant/ facility. Therefore, maintenance policy deals with asset performance, reliability, age, asset health, and asset life cycle while deciding its maintenance policy.

Maintenance budget is part of the annual budget plan of the plant/facility that is set aside in a financial year for performing maintenance activities of assets in the plant/facility. The budget is majorly planned in two categories, i.e., capital budget and revenue budget. The capital budget includes new procurement of assets, replacement of assets, and modernization of assets in obsolete equipment. The revenue includes a budget for routine maintenance, shutdown maintenance, preventive/ predictive maintenance, material procurement, and workforce contract.

Maintenance challenges (MC) are different from challenges in maintenance. These maintenance challenges are unforeseen and unexpected issues faced by maintenance managers/engineers. These maintenance challenges are specific to the industry, the process, and the equipment. Most of these issues reside with the maintenance managers/engineers as tacit knowledge. It is required to capture such experience and disseminate it among the units across the organization. Such knowledge sharing facilitates redesigning or selecting a suitable maintenance strategy and implementing it to adopt the best maintenance practices.

In this chapter, the concepts and theory behind asset maintenance management are described in detail. The recent developments in OR and MS and modern asset maintenance management systems enable to improve asset maintenance performance and achieve excellence. Increasing the efficiency during the asset life cycle period of an asset must be a key objective for asset maintenance practitioners. They must treat asset maintenance as a strategic function. Chapter 2 deals with an in-depth analysis of the asset maintenance contributing factors.

Key Takeaways

- Asset maintenance managers must think differently regarding asset failures and understand the diversity of asset maintenance issues. Merely adopting any asset maintenance strategy does not suffice, and there is a requirement to focus on various aspects of asset maintenance management and contributing factors of asset maintenance.
- An asset maintenance strategy is upkeeping the production plants/facilities and equipment in healthy condition. The strategy changes according to the nature of the production process and its operations. It is to formulate asset/equipment-wise best asset maintenance strategic plan in coordination with all the stakeholders.
- Asset maintenance is no more a functional activity. The organization's strategic move is to achieve its business goals in reducing asset maintenance costs and increasing production capacity.
- The asset life cycle represents the economic lifetime of the asset in an organization. The asset life cycle consists of the following stages: planning, procurement, commissioning and depreciation, operation and maintenance, and disposal.
- The operations-intensive organizations are described by costly specialized assets and rigorous environmental consideration. A key challenge in asset maintenance management is improving the asset maintenance performance continuously.
- The essentials of asset maintenance excellence are routine maintenance monitoring, identifying improvement areas, providing solutions to the identified improvement areas, and managing the asset maintenance manager's overall leadership.
- The modern maintenance management system must have a proactive and data-driven approach in designing intelligent maintenance systems. This kind of approach leads to a 35% reduction in the equipment/asset/plant's downtime.
- Modern maintenance management systems broadly cover e-maintenance applications such as Computerized Maintenance Management System and Enterprise Asset Management System. Intelligent automation systems like the Internet of Things and Cyber-Physical Systems enable asset maintenance management. The smart predictive technologies and tools developed based on these systems facilitate implementing the asset maintenance strategies effectively.
- The low MRO supply chain development affects various aspects of asset maintenance such as process, organization, technology, and inventory systems and the entire maintenance value chain.
- Maintenance, repair, and operations supply chain management are dealt with by two key departments in the organization like contracts and procurement and maintenance. The critical issue of the MRO supply chain is keeping a better inventory control mechanism.

- The major contributing factors in asset maintenance are maintenance tactics, reliability analysis, maintenance performance, planning and scheduling, materials management, organization/human resources, employee empowerment, information technology, maintenance policies/budget, and maintenance challenges.
- The effective management of the MRO supply chain can be accomplished by setting the basics right, i.e., identifying the problem areas and plan for improvement in those areas. Once the basics are set right, cloud-based software as a service (SaaS) can be used to manage master data of the MRO supply chain and establish adequate controls of inventory.
- Organizations aim to utilize mathematically based models for selecting the most appropriate asset maintenance strategic plan with both operations and maintenance departments' involvement. One such useful model is the Analytic Hierarchy Process (AHP). This method prioritizes the criteria for asset maintenance strategy selection and ranking among the selected asset maintenance strategies.
- The recent developments like Real-Time Location System (RTLS) in alliance with the Warehouse Management System (WMS) help quickly identify the required spare parts available within the organization and further even determine the availability of spare with the vendor warehouse.

References

Aberdeen. (2006). *Report on collaborative asset maintenance strategies.*

Afrinaldi, F., Tasman, A. M., Zhang, H. C., & Hasan, A. (2017). Minimizing economic and environmental impacts through an optimal preventive model and application. *Journal of Cleaner Production, 143*, 882–893.

Alsyouf, I. (2009). Maintenance practices in Swedish industries: Survey results. *International Journal of Production Economics, 12*(1), 212–223.

Bram, D. J., & Philip, A. S. (2020). A review on maintenance optimization. *European Journal of Operations Research, 285*, 805–824.

Choi, T. M., Wallace, S. W., & Wang, C. (2018). Big data analytics in operations management. *Production and Operations Management, 27*(10), 1868–1883.

Gits, C. W. (1994). Structuring maintenance control systems. *International Journal of Operations & Production Management, 14*(7), 5–17.

Ilangkumaran, M., & Kumaran, S. (2012). Application of hybrid VIKOR model in the selection of maintenance strategy. *International Journal of Information Systems and Supply Chain Management, 5*(2), 59–81.

Mobley, R. K. (2008). *Maintenance engineering handbook*. New York: McGraw Hill.

Salonen, A., & Deleryd, M. (2011). Cost of poor maintenance—A concept for maintenance improvement. *Journal of Quality in Maintenance Engineering, 17*(1), 63–73.

Velmurugan, R. S. (2016). *Doctoral Thesis. Maintenance strategy and practices at a large gas utility company in India—a case study approach*. Dehradun, India: University of Petroleum and Energy Studies.

Chapter 2
Asset Maintenance in Operations-Intensive Organizations

Abstract This chapter provides an overview of operations-intensive organizations. The deep insights and applications of the asset maintenance contributing factors during asset maintenance phases such as the initial phase, implementation phase, and assessment phase are explained based on the research study conducted among asset maintenance managers/engineers working in operations-intensive organizations. This approach adds value to asset maintenance concepts by capturing tacit knowledge available to the industry.

During the initial phase of asset maintenance, the key elements to be considered in planning and scheduling maintenance work are equipment history, work order with priority, preventive maintenance schedules, weekly work schedules, long-term/shutdown maintenance schedule, and work measurement backlog. A suitable Computerized Maintenance Management Software that integrates all these elements to maintain the maintenance scheduling and monitor work schedules is essential in asset maintenance. The importance of maintenance policy, allocation of budget resources, the role of materials management, and workforce resources are discussed in this chapter.

During the asset maintenance implementation phase, the various elements of maintenance tactics such as preventive maintenance, predictive maintenance, total maintenance hours, compliance of maintenance schedules, Original Equipment Manufacturer (OEM) recommendations, Reliability-Centered Maintenance, Total Productive Maintenance, proactive maintenance, and shutdown maintenance are also deliberated in detail in this chapter. Further, the first area of change toward employee motivation should be employee empowerment. Maintenance managers should provide directions and point out the desired results; empowering maintenance supervisors to achieve the desired results within their capacity is elaborated in the employee empowerment topic.

The key elements in evaluating maintenance performance such as maintenance cost, equipment history, training man-hours of maintenance staff, internal/industry norms, benchmarking measures and targets, and their role in formulating and selecting maintenance strategy are examined in this chapter.

Maintenance practitioners believe that the maintenance challenges are helping in formulating maintenance strategies for the company. Maintenance challenges are the

R. S. Velmurugan, T. Dhingra, *Asset Maintenance Management in Industry*, International Series in Operations Research & Management Science 310, https://doi.org/10.1007/978-3-030-74154-9_2

major issues faced by maintenance managers in their day-to-day asset maintenance function. Such issues are grouped into technical issues, contract issues, stakeholder issues, internal issues, and unexpected issues. Each of the asset maintenance contributing factors is supported by Industry Quotes in the Industry Quote Box.

2.1 Operations-Intensive Organizations: Overview

Operations-intensive organizations strive to manage their operations and manage capital assets effectively to ensure profitability. These organizations face several common challenges due to operating and maintaining expensive, huge, and remote assets. Such challenges are from optimizing the complex global supply chain scenarios to mitigating the impact of environmental issues. Poor management of the operational processes results in a significant cost increase that negatively impacts profitability. International and local laws to minimize the environmental impact are a significant concern for organizations. They are also under regulatory frameworks of the country and industry standards to maintain all stakeholders' health and safety, including protecting their workforce involved in the operations.

Operations-intensive organizations face several issues such as high capital investment, infrastructure bottlenecks, increasing raw material costs, low domestic/international demand, excess supply in the market, and interest rate of debt, which led to a slowdown in the business growth. To mitigate these issues, achieve the aimed GDP growth, and sustain the world market, the organizations require to take concrete steps. One such effort is pruning the operational and maintenance costs of the production. Therefore, the global best practices and strategies are to be implemented in operations and maintenance functions at an organizational level aligned with the industry's benchmarking.

The operations-intensive organizations work upon the following objectives tactfully with the help of asset maintenance as a primary support function:

- Monitoring of all assets/equipment at every stage of the life cycle
- Maximizing Return on Investment (ROI) by ensuring better performance of assets with less maintenance cost
- Improving asset utilization for maximum plant workload and further optimizing the operations and maintenance cost
- Deploying better solutions using Enterprise Resource Planning systems to manage and analyze data for providing preventive or predictive controls on health, safety, and the environment.

Operations-intensive organizations from the industries like automotive, mines, oil and gas, and chemicals recognize that asset maintenance is crucial to their organizations' successful operations. In summary, it emphasizes operational excellence, reduces the cost of asset maintenance, and ensures regulatory compliance on occupational health, safety, and the environment to drive profitable and sustainable growth of the organizations in the industry. In the present economic situation,

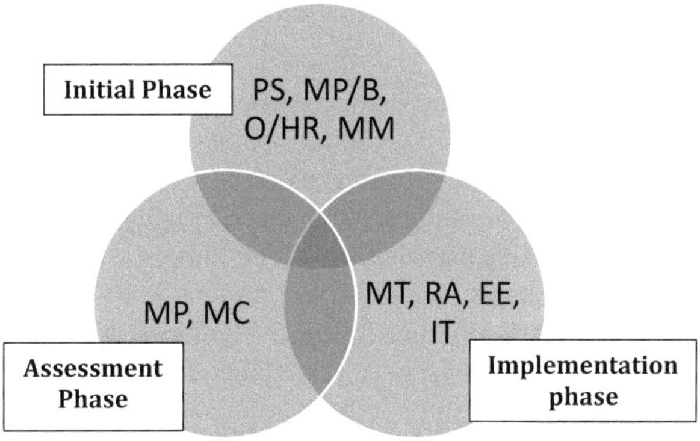

Fig. 2.1 Assessment maintenance phases

volatile prices and unstable customer demand, these organizations must make quick and difficult operational and business decisions. To remain competitive, organizations are examining new ways to get the most out of their assets, plan for continuous assets availability for 24/7 operations, and avoid unexpected failures.

The primary application of the asset maintenance contributing factors during asset maintenance phases is the initial phase, implementation phase, and assessment phase. During the initial phase of asset maintenance, planning and scheduling (PS), maintenance policies/budget (MP/B), materials management (MM), and organization/human resources (O/HR) play a crucial role. Similarly, during the implementation phase, the contributing factors are maintenance tactics (MT), reliability analysis (RA), employee empowerment (EE), and Information Technology (IT). The assessment phase's major contributing factors are maintenance performance (MP) and maintenance challenges (MC). Working on these elements overlappingly during the entire asset maintenance management cycle is important, and the elements are depicted in Fig. 2.1.

Asset maintenance function of any plant/facility in terms of asset maintenance contributing factors provides insights on maintenance activities and evaluates such a plant/facility's asset healthiness. This chapter describes critical elements for each of the asset maintenance contributing factors in operations-intensive organizations. This description is based on research study and experience from maintenance practitioners in the industry.

2.2 Asset Maintenance: Initial Phase

The initial phase of maintenance is an important phase since the planning for the material, workforce, and maintenance jobs is carried out. The famous proverb says, "Well begun is half done." As told in the saying, the excellent planning leads to maintenance excellence. The cost allocation, i.e., budgeting and setting benchmarks for maintenance performance, is also built-in in this stage. The contributing factors in the asset maintenance initial phase are discussed in this section.

2.2.1 Planning and Scheduling

In operations-intensive organizations, appropriate planning of operations and maintenance activities at an organizational level is the starting point of asset maintenance management. Such planning must be integrated with all business processes for seamless implementation. Computerized manufacturing concepts offer Smart Solutions for effective planning and scheduling of asset maintenance activities in modern asset maintenance. Additionally, predictive maintenance inputs of an asset/equipment are also incorporated proactively during the planning and scheduling stage to ensure the maximum asset availability for operational profits. For example, asset maintenance efficiency can be maximized using Total Productive Maintenance tools by facilitating optimum utilization of equipment in operations/manufacturing in the automobile industry.

Asset maintenance planning identifies the essential requirements such as workforce, material, prerequisite, and Original Equipment Manufacturer (OEM)/external support requirements to perform a maintenance job. Once the planning is done, then the necessary preparatory activities to fulfill the essential requirements must be initiated and completed before starting the job. The preparatory activities are placing work orders for hiring workforce/labor and purchasing orders for spares and consumable material requirements, preparing worksheets and drawings, and collecting all datasets before scheduling the jobs. Good planning is a necessity for accurate scheduling.

Asset maintenance scheduling is a critical factor in executing the asset maintenance plan and achieving operational competence in the organization. In scheduling, a list of jobs is prepared with priority, timelines, and sequence as per the critical path of execution by allotting the required resources. The jobs are scheduled in a phased sequence manner considering the operational requirements with the specific time for execution of the job. The asset maintenance schedule must be built-in with the factor for job monitoring, controlling, and reporting the progress. The key consideration in scheduling is maximum work with the available resources, prioritizing schedules, and minimizing contract and external (OEM) resources by utilizing inside expertise.

The key elements (Fig. 2.2) required to be considered in asset maintenance planning and scheduling are equipment history, work order with priority, preventive maintenance schedules, weekly work schedules, long-term or shutdown maintenance schedules, and backlog work measurement (Velmurugan, 2016). An appropriate Computerized Maintenance Management System (CMMS) is essential to maintain the maintenance activities' scheduling and monitor such schedules for timely completion.

Equipment history must include equipment specification data and maintenance works performed since commissioning of the equipment, PM schedules as per OEM recommendations, minor/major modifications carried out, and equipment availability. Asset breakdown reports like incident analysis reports, root cause analysis, and other similar analysis reports must be part of equipment history available in a Computerized Maintenance Management System.

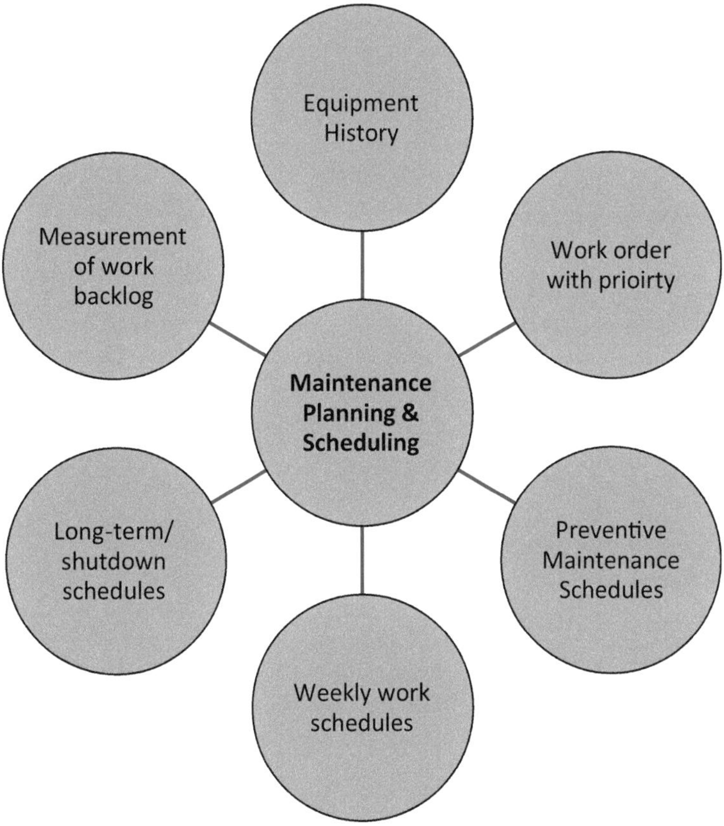

Fig. 2.2 Maintenance planning and scheduling

Work order with priority is an element that requires translating preventive maintenance schedules into the job. The PM schedule execution requires clearance from operations and safety personnel of the plant/facility. The maintenance managers/engineers use work order monitoring to plan and document the job carried out on specific equipment within an organization. Work order systems are a must because of the amount of work the maintenance team needs to monitor and control. The work order systems are automated with the use of maintenance application software and can be monitored centrally.

Preventive maintenance schedules are developed in line with Original Equipment Manufacturers' recommendations, ISO or any other industry standard requirements, standard operating procedures, maintenance manuals, equipment types, and the maintenance team's experience in operations-intensive organizations. **Weekly work schedule** availability based on the PM schedule must take clearance from all the stakeholders to execute the required maintenance without any hurdle as per plan.

Long-term/shutdown maintenance Planning depends on types of assets such as compressors, turbines, pumps, Computer Numerical Control (CNC) machines, robotic assembly, and market requirements. The *centralized procurement method* could be beneficial for multifacilities across the organization to have comprehensive planning for long-term planning. Shutdown maintenance requires meticulous planning in consultation with all stakeholders, especially with the Original Equipment Manufacturers of assets/equipment and process licensors. Shutdown planning must also include the plant/facility's prestart activities plan and the start-up procedure of operations.

Industry Quote Box 2.1: Planning and Scheduling

- Short-term maintenance planning includes preventive maintenance (PM) schedules. These schedules are monitored every month.
- PM schedules and frequencies are planned based on asset maintenance team's experience and recommendations from the equipment manufacturer.
- PM schedules' frequency varies with the types of equipment. The frequencies are monthly, quarterly, or yearly basis.
- We consider equipment history and age of the equipment while preparing maintenance schedules.
- Shutdown maintenance planning is a major activity in our petrochemical facility, and it depends on types, criticality, nature of operations of an asset, or equipment.
- SAP system is available in our organization, and SAP plant maintenance module is used to monitor the completion of maintenance scheduling.
- SAP PM module indicates the deviation in completion of the maintenance schedules. The deviation is informed to operations in-charge, and we take necessary approval of such deviations. The records are maintained as per maintenance policy.
- Our challenge in backlog maintenance jobs overburden the planned maintenance jobs. The same is tackled with additional resource allocation of manpower and material.
- In case of breakdown of an equipment, root cause analysis is carried out in our organization, and incident analysis reports are prepared, and further maintenance schedules are modified, in case of any requirement.
- In pipeline transmission of our organization, standby equipment is available in critical operations. Therefore, shutdown maintenance is not much critical like petrochemical operations.

The amount of **backlog work orders** is an important maintenance metric. This metric helps work type and coverage, schedule completion time, compliance, and backlog work hours. From the industry data as an example, it is also evident that in pipeline maintenance activities of gas transmission and distribution companies, maintenance schedules cannot be carried out during crop or rainy season, so that the same is taken up during the non-crop season. However, such backlog jobs are taken up appropriately by the organization during the opportunity available subsequently if any additional labor requirements, spares, and other resources are also allocated for timely completion of the backlog jobs (Velmurugan, 2016).

2.2.2 Maintenance Policies/Budget

Maintenance policy and maintenance budget are essential contributing factors in asset maintenance management. The key elements (Fig. 2.3) for asset maintenance are maintenance mission and objectives, integration of maintenance plan and business plan, maintenance policies, and maintenance budget (Velmurugan, 2016).

Maintenance mission/vision is an essential factor for an operations-intensive organization to maintain assets/equipment in a healthy state. The significant aspects of maintenance policy must satisfy the following:

- Asset/equipment is always in hot standby condition, i.e., the asset can be taken in-line immediately in case of an operational requirement.
- Asset/equipment is always in quality tested condition to enable quality production and meet the competitive advantage. The frequent breakdown and start-up or

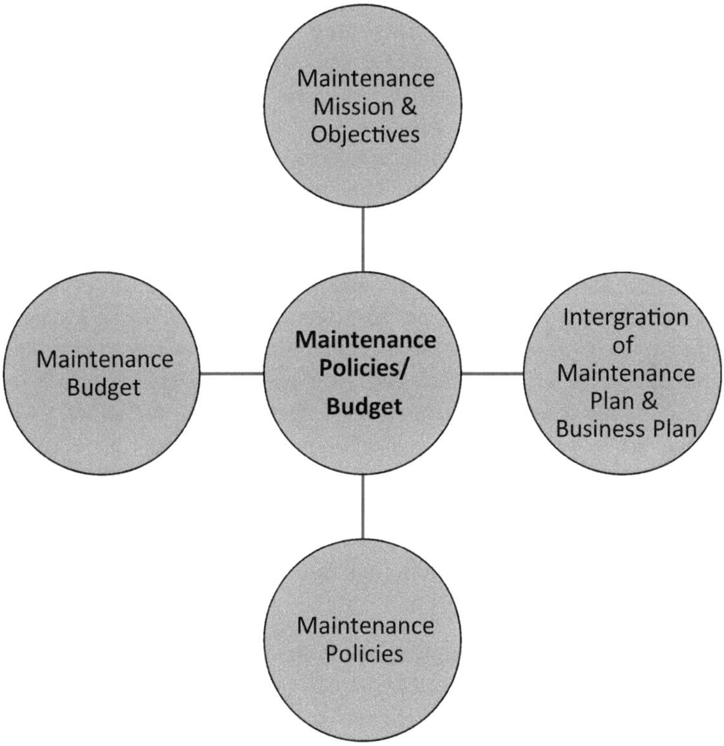

Fig. 2.3 Maintenance policies/budget

product changeover can impact product quality. The maintenance policy should be aligned with such production/operational requirements.
- The major breakdowns of assets/equipment should be avoided. The major breakdown leads to a reduction in product inventory and, ultimately, the company's market share loss.
- Asset maintenance costs are always monitored and controlled.
- Maintenance obsolescence should be included as a checkpoint.
- Maintenance audits should also be part of the maintenance policy.

The effectiveness of asset maintenance strategic plan implementation depends on maintenance budgeting. The **maintenance budget** estimation is prepared based on annual maintenance jobs and the long-term jobs planned and scheduled. The maintenance budget cost includes the workforce cost, i.e., permanent employees of the organizations and the contract employees deployed, cost of spares, and materials required to carry out the maintenance jobs. The primary process plants operating on a 24/7 basis require a separate budget for executing annual turnaround jobs.

The senior management always treats the asset maintenance budget as a necessary cost to the company and does not care about the relation between the asset maintenance expenses and the production output (Mobley, 2008). Such treatment exists in large process/manufacturing plants or an individual manufacturing unit in operations-intensive organizations.

The asset maintenance budget is always aligned with the corporate guidelines for preparing capital expenditures or operational expenditures. The current financial year's budget is reviewed and revised every year based on the existing resources (workforce, material, and other resources) for asset maintenance purposes. An estimated budget for asset maintenance is prepared for the next financial year to accomplish planned maintenance jobs. This budgeting cycle repeats every year.

Integration of maintenance plan and business plan is a critical element in operations-intensive organizations. Such kind of integration strategy should be a proactive action for the long-term success of the business. Asset maintenance managers must leverage technological developments such as network connectivity, advanced software platforms, and intelligent automation systems to accomplish this integration across the organization. The asset maintenance information can flow from the plant to the corporate level in a cost-effective manner. Further, CMMS and ERP systems are to be integrated for facilitating data exchange between corporate and plant asset maintenance. For example, a shutdown maintenance plan is essentially based on annual sales targets and operational planning. Therefore, operations-intensive organizations must reorient the asset maintenance functional strategy across the available plant-wide systems and business processes.

Industry Quote Box 2.2: Maintenance Policies/Maintenance Budget

- Maintenance policy is available in our organization. The maintenance policy has been prepared in alignment with the organization's mission

(continued)

Industry Quote Box 2.2 (continued)

and vision. Our maintenance activities are planned and scheduled as per the guidelines given in the maintenance policy. One example is that the deviation in maintenance schedules is to be approved by the plant in-charge. A well-written format is provided in the maintenance policy.

- Maintenance policy is a guideline for execution of asset maintenance plan. Each unit maintenance head plans the maintenance activities as per the policy.
- Maintenance mission and vision are the driving forces of the maintenance policy, and the management approach for developing asset maintenance strategy is available in our organization's maintenance policy.
- Maintenance policy in our organization broadly covers the availability of asset to operations, healthiness of equipment and systems, systematic and structured approach to implementing the preventive maintenance schedules, and predictive maintenance activities.
- Budgeting is a major annual activity in our organization. The key considerations in our plant while preparing maintenance budget are stock levels of spares and replenishment of the spares during the next financial year, major or minor modifications planned, vendor service requirements, etc.
- Our key consideration in maintenance budget is the ratio between preventive maintenance and predictive maintenance activities. If we include procurement of predictive maintenance-based systems like intelligent hardware and software, the budget estimation would also increase.
- In our company, the annual budget is prepared compared to actual expenditure of last year vs. the budget allocation. The major modifications and asset replacements are part of the capital budget. Spares, services from OEM, and contract workforce are part of the revenue budget.

2.2.3 Materials Management

Materials management comprises the following processes: identify spares/consumables, purchase the identified spares, manage the spare inventory at stores and shop floor, issue to the concerned department based on the job requirement, and repeat the process on the low-stock inventory of identified spares/consumables. As part of the departmental activities, these processes come under procurement, stores management, inventory control, and asset maintenance functions. The coordinated activities among all these functions are the deciding factors in implementing asset maintenance strategic plans in operations-intensive organizations.

Even the asset maintenance strategy is well planned, in case of nonavailability of spares that leads to catastrophic conditions to the asset maintenance performance. Therefore, the successful implementation of the asset maintenance plan entirely

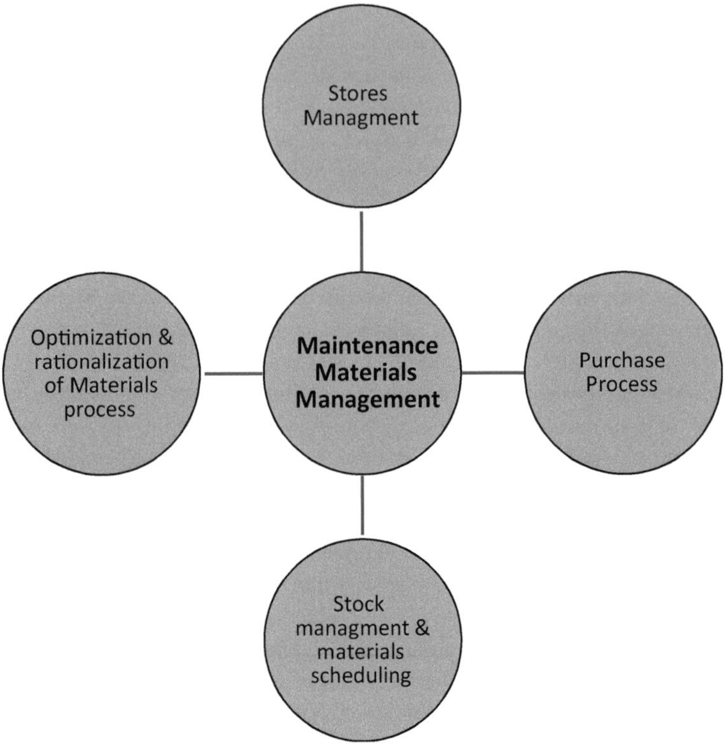

Fig. 2.4 Maintenance materials management

depends on the asset maintenance team's competent materials management in collaboration with the procurement team. This challenge can be tackled with appropriate computerized materials management system integrated with the maintenance management system.

The key elements (Fig. 2.4) in materials management related to asset maintenance are purchase process, stock management and materials scheduling, optimization and rationalization of material process, and stores management (Velmurugan, 2016).

The purchase process is normally robust in any organization. However, the purchase process, which can handle an emergency, is more effective in the maintenance material process. The emergency purchase procedures facilitate asset maintenance managers to tackle unexpected breakdowns of assets, and while preparing maintenance budgets, provisions for emergency procurement can be kept for mitigating such emergencies. Most operations-intensive organizations use the **centralized procurement** concept to ensure just-in-time spare availability, maintain optimum spares across various plants/facilities, and plan for shutdown maintenance. The purchase process of materials depends on types of spares such as critical or noncritical, utilization pattern of spares/consumables, criticality of assets/equipment, procurement lead time, and nature of spare procurement such as OEM spares or available in the open market.

Stock management and materials scheduling are important components in materials management related to asset maintenance management. In this process, it is apparent from the industrial data that stock management practices such as categorization of spares, standard procurement procedures, well-documented stores manual, and contract procedures with a delegation of powers are available with the organizations. The categorization of spares is critical spares and normal spares. Critical spares are spares of critical equipment/asset, spares with high procurement lead time, and high-value spares. Normal spares are consumables, spares with low procurement lead time, and low-value spares.

Optimization and rationalization of the material process are majorly involved in inventory analysis and control. Based on asset/equipment historical data, stock position of spares, physical verification of nonmoving stock, and inventory analysis are carried out in the organizations annually. The accurate inventory analysis can manage and control the spares of assets/equipment prudently. For better inventory analysis and control, the Enterprise Resource Planning system and Computerized Maintenance Management System are to be integrated for aligning both materials and asset maintenance plans. These data are further to be monitored by the procurement department and stores. Notification can be generated regarding spare inventory to concerned asset maintenance managers regularly.

Industry Quote Box 2.3: Materials Management

- Long-term maintenance planning and procurement cost are the factors taken in our maintenance activities. Accordingly, centralized procurement policy is followed in our organization for several years.
- Procurement of spares consumes longer lead time. Therefore, we plan spare procurement based on types of the spare, consumption pattern, critical nature of spares, etc.
- We categorize spares like compressor rotor assembly as insurance or critical spares; such spares requirement rarely depends on the major failure of equipment, but it is essential to keep such spare items.
- We tried with indigenization of spares during the retrofitting of one of our equipment, but we could not succeed since replaced spare failed within a short period. Therefore, our maintenance policy spells out that the spares must be procured from OEMs only.
- Efficient spares management leads to higher plant availability. Therefore, our top management reviews spare availability quarterly in our organization.
- Spare inventory are periodically reviewed and controlled based on equipment/asset history. We follow the checks and balances like annual

(continued)

Industry Quote Box 2.3 (continued)
verification of nonmoving spare items and overall stock verification to ensure inventory control.
- SAP materials management module is used for inventory management of all materials including spares management in our organization.

Stores management deals with the elements such as spare availability, spare identification with traceability, an inspection of incoming materials, storage of material in a specific and categorized location, bin management, and regular review of material at stores. Such material review includes unused spares for a long time, regular spares, inventory of consumables, and updating data in the ERP system. The well-documented stores manual facilitates standard work procedures in materials stores in operations-intensive organizations.

2.2.4 *Organization/Human Resources*

Organization and human resources are essential asset maintenance contributing factors for implementing asset maintenance strategy and asset maintenance performance. The manufacturing strategy and business philosophy are continually changing due to the VUCA (volatility, uncertainty, complexity and ambiguity) business environment. In this environment, organizations are adopting the philosophy to outsource asset maintenance-related jobs. This approach facilitates organizations that they do not require to create permanent infrastructure or assets. The asset maintenance team in any organization must be capable of handling multifunctional activities in asset maintenance.

Knowledge and skills are important for any asset maintenance team member to perform the maintenance job efficiently. Asset maintenance-related training programs help maintenance engineers/technicians acquire knowledge on asset maintenance systems, and additionally, the skills (on-the-job application of knowledge) earned by the individual improve the maintenance job performance. Therefore, the maintenance engineer/technician's role portrays a critical role in accomplishing maintenance excellence metrics. The operations-intensive organizations must consider training and development as one of the organization's performing areas, and it is not the subject of cost-pruning measures.

HR planning for asset maintenance organizational structure depends on the various determinants such as job description, job responsibilities, organization structure, a span of control, operational requirements, and benchmarked industry practices. These are to be finalized considering the asset maintenance philosophy at an organizational level. The philosophies are centralized vs. decentralized asset maintenance and maintenance job execution by in-house workforce vs. outsourced workforce.

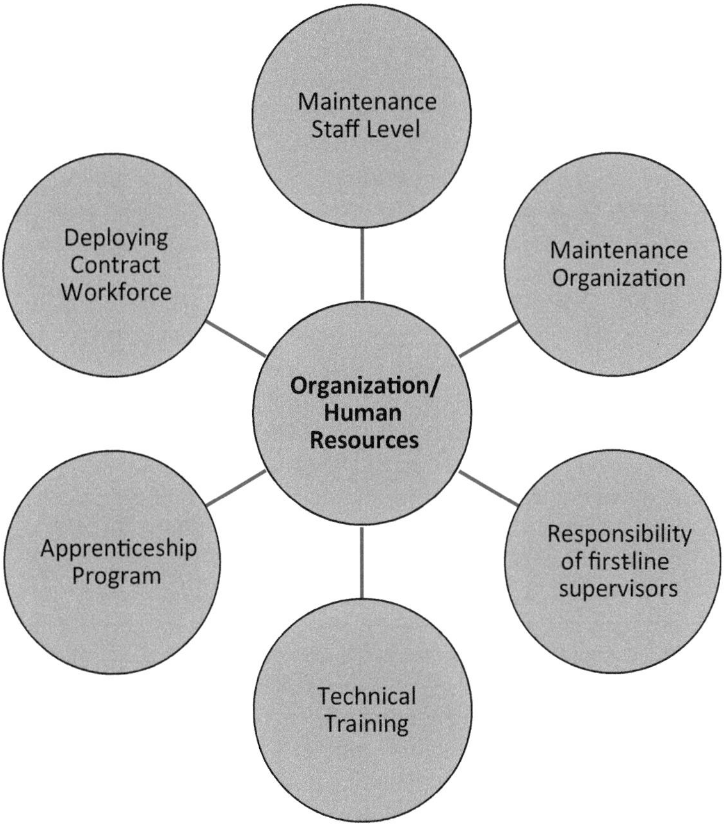

Fig. 2.5 Maintenance organization/human resources

The asset maintenance managers should be capable of planning the maintenance worker division to perform asset maintenance jobs and achieve the common asset maintenance objectives in a systematic approach. Therefore, an appropriate selection of asset maintenance teams with the specialized capabilities in asset maintenance function supported by training and development interventions is the current business need of operations-intensive organization.

The key components (Fig. 2.5) that play a role in asset maintenance are maintenance staff level, maintenance organization, the responsibility of first-line supervisors and adequacy of support staff, regular technical training, apprenticeship program, and contractors (Velmurugan, 2016).

Maintenance staff level is an essential element in some operations-intensive organizations like oil and gas processing plants due to a hazardous environment, wherein the workforce is deployed on a 24/7 basis. The maintenance staff should have a suitable technical qualification(s) with sufficient technical knowledge and are professionally trained in the area where they are placed.

Maintenance organization explains that the facility should be well structured, covering all the maintenance disciplines such as electrical, mechanical, civil, and instrumentation in the context of asset maintenance of operations-intensive organizations. The maintenance job requires covering the whole spectrum of jobs like equipment troubleshooting, instrument calibration, maintenance of valves, maintenance of field instruments, and operations of compressors, pumps, and CNC machines. Therefore, asset maintenance engineers/technicians must be skilled in specific maintenance jobs and trained on asset/equipment maintenance from the Original Equipment Manufacturer.

Responsibility of first-line supervisors is important, and the job description of these supervisors must be clearly defined since the day-to-day jobs are executed by them firsthand. The concerned departments' supervisors from electrical, mechanical, instrumentation, and civil departments update CMMS's maintenance job order status after performing the specific maintenance job. Their inputs are crucial in measuring asset maintenance performance in operations-intensive organizations.

Imparting **technical training** to the maintenance team for all the assets/equipment and associated control systems is essential for quick troubleshooting of failures and restarting the equipment within the shortest time possible to ensure continuous or 24/7 operations. Further, OEM expertise requirement can also be curtailed by providing more advanced training on the specific equipment/asset, which led to maintenance cost saving and increase in critical equipment availability.

The research study outcome shows that operations-intensive organizations are conducting the **apprenticeship program** as per industry data and statutory requirements of the country where they are located.

Deploying contract workforce, including proficient workforce from OEMs on a need basis, is the asset maintenance management's backbone. In consultation with the contracts and procurement department, the asset maintenance department accomplishes contract workforce deployment as per organizational policies. Overhauling of critical equipment is planned during the planned plant shutdown. The additional contract workforce required for the shutdown jobs is hired through a separate contract by the contracts department. The asset maintenance manager is responsible for executing the job during the shutdown after awarding the contract. For day-to-day asset maintenance jobs, the workforce is used from the existing annual maintenance contracts.

Industry Quote Box 2.4: Organization/Human Resources

- In our company, the maintenance organization includes the departments like mechanical, electrical, instrumentation, telecommunications, and civil. We all are reporting to the maintenance head, senior-level executive.
- Technicians and engineers are coming in shifts in process plant operations of our organizations since the plant is operating on a 24/7 basis.

(continued)

Industry Quote Box 2.4 (continued)

- Our company follows well-structured maintenance organization at a unit level. There is no corporate maintenance department as the case in some of the other companies.
- We have multiple process facilities across the country, even though a dedicated team is available at the unit level; corporate maintenance department takes care of centralized procurement, vets major modifications, formulates and revises the maintenance policy, and coordinates among the process facilities to achieve common objectives of the organization.
- Our maintenance activities require skilled manpower to work on specific equipment or operations, and in some of the remote locations, the workforce is trained to take up jobs in case of absence of deployed manpower.
- Frontline supervisors are responsible for executing planned maintenance activities in our plant. Department-wise (electrical, mechanical, instrumentation, etc.) heads are responsible for updating the maintenance activities completion in maintenance management systems.
- As per our maintenance policy, we are responsible for providing technical and safety training to outsource workforce deployed at our operating locations before assigning jobs to them. Refresher training is conducted every quarter.
- Some of our remote stations are uncrewed, and only security personnel is deployed for safeguarding the assets. We provide regular training quarterly, and documentation of such training is done to show to external auditors.
- The head of the maintenance department is empowered to hire experts for performing maintenance jobs in critical equipment at our organization.

2.3 Asset Maintenance: Implementation Phase

Implementation of asset maintenance management is the efficient delivery of the asset maintenance strategic plan. In this phase, maintenance tactics or methods used suitable to the type of process plant or facility are critical in implementing the best maintenance practices. Reliability analysis of the asset maintenance implementation ensures the correct execution of asset maintenance processes. The complete asset maintenance implementation supported by information technology (IT) infrastructure and employee empowerment enables successful asset maintenance management.

2.3.1 Maintenance Tactics

Maintenance tactics are one of the major contributing factors in asset maintenance management. The various elements (Fig. 2.6) of maintenance tactics are preventive maintenance, predictive maintenance, Original Equipment Manufacturer recommendations, Reliability-Centered Maintenance (RCM), Total Productive Maintenance (TPM), and shutdown maintenance (Velmurugan, 2016).

Preventive maintenance (PM) is the maintenance of an asset/equipment of plant/facility as per the planned schedule of the equipment for maintaining the healthiness of the asset during its life cycle period by preventing excess damage and extending the asset life cycle period by performing maintenance activities such as servicing, replacements of recommended spare parts, calibration of instruments, and lubrication of equipment.

In addition to the conventional preventive maintenance models, preventive maintenance uses mathematical models to minimize the risk of asset/equipment failure

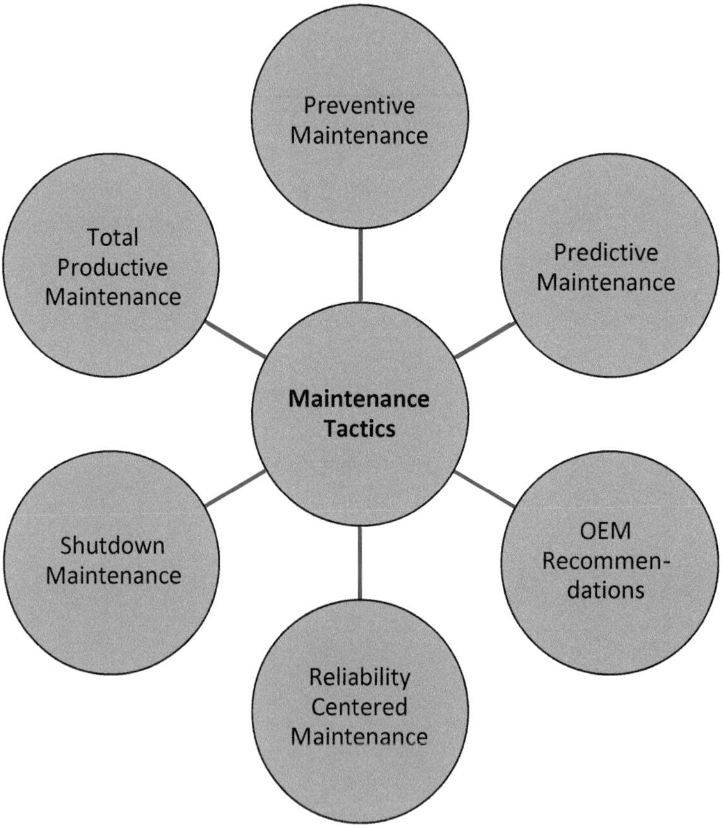

Fig. 2.6 Maintenance tactics

and maintain reliable equipment availability. There are also preventive maintenance models using fuzzy Bayesian tools for equipment maintenance. Therefore, developing an asset maintenance strategy using mathematical models is a valuable addition to asset maintenance management.

Predictive maintenance (PM) is a maintenance method that identifies the early failures before such failures become potential reasons for equipment/plant shutdown. The asset/equipment healthiness is monitored on a real-time basis from the datasets obtained from the devices fitted on various equipment parts. These datasets and the field measurements using portable devices are processed through CMMS to initiate maintenance actions. Preventive maintenance can be planned accurately based on the predictive maintenance inputs. The traditional predictive maintenance method is the physical inspection of an asset/equipment at the site and observes its condition by listening to the equipment sounds, measuring the temperature, and monitoring field conditions.

Operations-intensive organizations are already adopting the predictive maintenance models. Predictive maintenance requires additional infrastructure to be installed as part of the equipment to measure the parameters such as vibration, temperature, pressure, flow, and speed. Such infrastructure addition increases the maintenance cost of the organization. Predictive maintenance actions are initiated based on the trends and levels of measuring parameters mentioned above. Such actions include failure prediction before it occurs, root cause identification of the failures, and maintenance action to avoid the predicted failures before it causes severe equipment/asset damage.

Preventive maintenance and predictive maintenance are majorly used in operations-intensive organizations. These maintenance methods are planned based on OEM recommendations, internal norms, industry standards, types of equipment/asset, asset/equipment historical data, and previous experience of asset maintenance personnel.

Industry Quote Box 2.5 (a): Preventive Maintenance

- We majorly follow preventive and predictive maintenance in our process plant. All the assets are covered under preventive maintenance, and preventive maintenance schedules are uploaded in our SAP PM module, and jobs are executed and monitored through this system.
- Preventive maintenance schedules are part of our Quality Management Systems (ISO) in our company. These schedules are prepared in-line with OEM recommendations and our previous experience.
- Industry standards and company norms influence our preventive maintenance strategies, and accordingly, the preventive maintenance schedules and frequencies are decided in our unit.
- We review preventive maintenance schedules on an annual basis. The PM schedules are revised on the elements like aging equipment and failure

(continued)

Industry Quote Box 2.5 (continued)

history. Even in some cases, we increased PM frequency from 6 months to 1 year.

- One of our maintenance performance metrics is preventive maintenance cost. Therefore, our process plant's PM schedule is reviewed based on PM cost also.

Preventive maintenance schedules are executed with the appropriate resources as per the asset maintenance plan. Such schedules are reviewed annually based on Quality Management System (QMS) manuals and procedures of the asset maintenance process at an organizational level. Preventive maintenance schedule completion is confirmed through QMS audits from both internal teams and external experts. Predictive maintenance actions are integrated with CMMS for initiating asset maintenance job orders to rectify the predicted failures of the assets by the asset maintenance team.

Industry Quote Box 2.5 (b): Predictive Maintenance and Other Tactics

- Predictive maintenance becomes an integral part of our maintenance policy. All the critical equipment like compressors, turbines, and furnaces are integrated into the predictive maintenance environment. Now, we are also revamping our existing predictive maintenance systems to intelligent systems for easy diagnostics of failures.
- The major predictive maintenance of equipment is vibration monitoring and analysis in our facility. The same is taken care of by inspection department in our organization.
- Decision of overhaul of critical equipment is taken based on predictive maintenance analysis and OEM recommendations. Sometimes, we decide to postpone the equipment overhaul duration to increase the equipment availability to catch up with the production targets.
- We do not follow any Reliability-Centered Maintenance techniques in our organization. However, incident analysis reports are used to improve the maintenance performance, which helps to improve reliability.
- Standby equipment strategy in critical operations helps to avoid long shutdown maintenance in our facility.
- We maintain reliability records like equipment history, equipment breakdown records, and incident analysis reports for improving reliability in maintenance.
- Emergency maintenance is part of the maintenance tactics in our unit. Emergency preparedness plan is available and shared among all the stakeholders. Emergency response vehicle (ERV) is kept ready at our unit for handling any emergency arises.

Reliability-Centered Maintenance (RCM) is a maintenance approach that ranks assets/equipment in their criticality and availability. RCM is a comprehensive approach by balancing all other asset maintenance tactics like preventive, reactive, predictive, and proactive to maximize the equipment/asset availability and minimize the asset maintenance cost. Impact on plant availability and capacity is assessed with the RCM concepts like asset/equipment prioritization and failure modes, asset maintenance resources augmentation, implementation of asset maintenance technologies, core competency development in asset maintenance function, asset maintenance mission for operational capacity development solutions, dissemination of impact of asset maintenance decisions, and the formation and implementation of asset maintenance metrics aligned with corporate business objectives.

Total Productive Maintenance (TPM) is a maintenance tactic that integrates the manufacturing process with asset maintenance. The key objective of TPM is to reduce the production loss due to asset maintenance, i.e., maximizing asset/equipment availability with no breakdown. Asset maintenance is always regarded as separate from the manufacturing process. Now, the operations-intensive organizations treat asset maintenance as part of the production process. TPM application eliminates the waste from the manufacturing process by reducing the production loss due to asset downtime. Reduction in asset rework and maximization of equipment availability contribute to value addition in asset maintenance management with minimum asset maintenance cost.

OEM recommendations are standard operating procedures (SOPs) and equipment maintenance procedures recommended by the Original Equipment Manufacturer of equipment under maintenance. The recommendations of OEM are available in the operations and maintenance manuals supplied by OEM along with the equipment. Such OEM recommendations must be strictly followed to ensure the warranty of the equipment. In case of any deviation, OEM does not warrant the equipment's life.

Shutdown maintenance is the maintenance approach in which the maintenance jobs are taken up during annual maintenance shutdown (or) during the plant shutdown specifically taken up for executing the asset maintenance of the critical assets/equipment as per the long-term asset maintenance plan. Shutdown maintenance may be of duration from few days to few weeks. The resources required are workforce, material, and other related resources, which are planned well in advance for shutdown maintenance to ensure the timely completion of the shutdown maintenance plan.

2.3.2 Reliability Analysis

The primary objective of asset maintenance management is maximizing the availability of assets for operations and plant uptime. Reliability analysis is an important asset maintenance approach, and the application of its techniques and tools aids in identifying the root cause of the failures and suggests options to rectify them. The

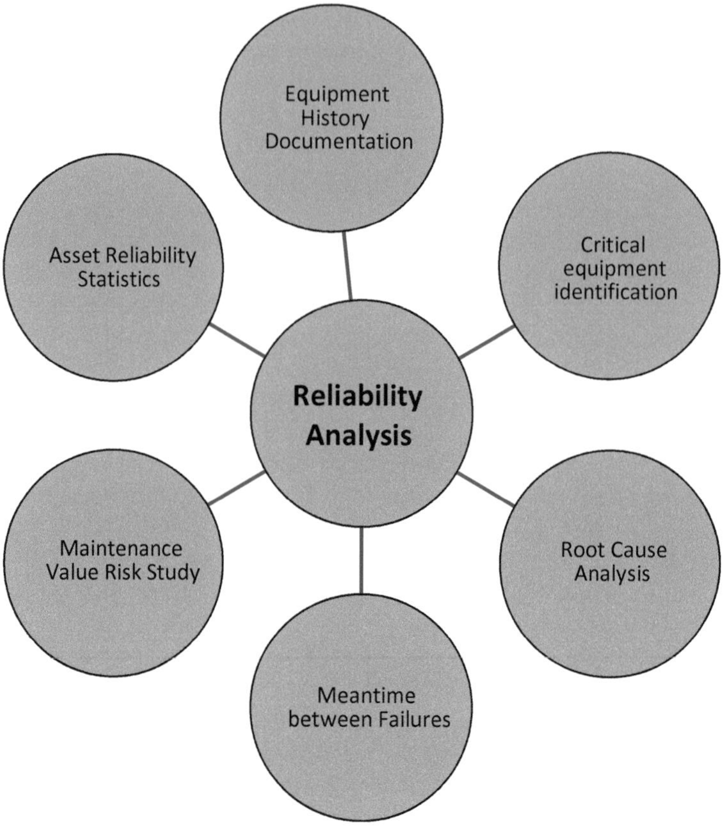

Fig. 2.7 Reliability analysis

key components (Fig. 2.7) related to reliability analysis are equipment history documentation, critical equipment identification, root cause analysis (RCA), mean time between failures (MTBF), maintenance value risk study, and asset reliability statistics (Velmurugan, 2016).

Documentation of asset/equipment history is known as **equipment history documentation**, and keeping equipment history records facilitates the reliability analysis. The equipment history records must contain the following as equipment details such as serial number, tag number, equipment specifications, asset life cycle duration, installation and commissioning data, details of equipment breakdown, reasons for failure, details of maintenance work carried out to rectify the equipment and expenditure incurred, and root cause of the failure. Breakdown history enables developing revised preventive maintenance schedules and spares replenishment, facilitates inspection of equipment during the annual shutdown to understand the failure rates and nature of failures, and incorporates necessary modifications in the equipment or process for avoiding repeated failures of equipment. *Root cause analysis* and equipment history learnings can be used to impart training to maintenance personnel, operations personnel, and contract employees responsible for equipment maintenance.

Critical equipment identification is an essential element in reliability analysis. The equipment's criticality is assessed from the organization's business impact in case of failure of the equipment. The equipment criticality does not depend on number of times the failure occurs, and it can be ranked with a range from one to ten for identifying the impact of the equipment criticalness. Score ten corresponds to the highest criticality nature, and score one corresponds to the equipment's lowest criticality nature. The equipment types, recommendations by OEM, and process requirements decide the equipment criticality. The maintenance schedules are planned according to the criticality of the equipment.

Industry Quote Box 2.6: Reliability Analysis

- We maintain the following records like equipment availability, breakdown details, standby equipment maintenance records, incident analysis reports, and root cause findings report for carrying out reliability analysis of a particular equipment.
- In our facility, we do systematic reliability analysis using asset maintenance system, and equipment history available in this system facilitates to make suitable revisions in our maintenance schedules.
- We use SAP plant maintenance module to carry out equipment reliability analysis, planning, and scheduling of maintenance schedules.
- Even though we apply few reliability analysis metrics, there is a requirement of training on RCM-based analysis to our maintenance personnel especially covering the aspects of quantitative methods in RCM.
- The availability of our pipeline transmission systems is 99.9%. Therefore, predictive maintenance is itself enough for ensuring the reliability of the system.

Root cause analysis (RCA) is an important technique that uses various tools and models to evaluate equipment/asset failure. This technique carries out a complete analysis to dig out the root cause of the failure. The identified root cause of nonconformance in the system is removed by improving the processes. The Reliability-Centered Maintenance (RCM) approach provides a detailed analysis methodology for the RCA process. An implementation of the RCM maintenance approach requires the revisiting of asset maintenance activities periodically. The failure causes and failure modes arrived through RCM methodology are used to improve and adjust the preventive maintenance schedules. Integration of RCM processes with CMMS/EAM provides more extensive benefits to operations-intensive organizations.

Mean time between failures (MTBF) is a reliability metric used to measure the number of failures per million hours of an asset/equipment/product. MTBF measure is used to assess the asset life cycle period, and asset maintenance managers can make asset maintenance interventions based on determining an asset/equipment. MTBF is proportional to the asset capacity threshold of an equipment/asset. The other related metrics are *Mean Time To Repair* (MTTR), *Mean Time Between Replacement* (MTBF), and *Mean Time Between Maintenance* (MTBM). MTTR is the average time to rectify the failure of equipment and make it ready for operations. MTBF is the average time of the equipment available for operations between replacements of equipment's parts due to maintenance or any other activities. MTBM is the average time between proactive (preventive or predictive) maintenance and reactive (breakdown or corrective) maintenance.

Maintenance value risk study is an element of reliability analysis in asset maintenance management. Risk can be calculated based on the below definition:

$$\text{Risk} = [\text{Probability of an event happening} * \text{Consequence of the event}].$$

In asset maintenance, the risk means deviation from the normal operating conditions and equipment parameters. Risk study of the maintenance values is a systematic process to identify the threat and estimate asset failures due to the risk. Such a study uses asset/equipment historical data, empirical analysis, stakeholders' views, and asset maintenance managers' experiences. Maintenance value risk study uses qualitative and quantitative techniques, and this study is guided by API 380 standard, i.e., Risk-Based Inspection in the industry.

Asset reliability statistics are based on reliability analysis, which observes the field data and creates mathematical models by inputting the industry's asset maintenance manager's experience. The well-developed models help understand the asset/equipment health parameters and interpret its performance behavior to minimize the operational costs during the asset life cycle period. Quantitative analysis methods such as life data analysis (Weibull analysis), dependability growth analysis, asset life cycle stress analysis, fault tree analysis, system modeling using reliability block diagrams, simulation, experimental designs, and other consistency predictions from the industry standards play a crucial role in asset reliability analysis. All these approaches are also process related with asset maintenance management.

2.3.3 Employee Empowerment

Employee empowerment is the best approach to keep employees motivated in organizations. Traditionally, asset maintenance planning and controlling activities are managed by the asset maintenance managers in the industry. In operations-intensive organizations, to improve operational and maintenance efficiency, the asset maintenance managers have delegated micro detailing of asset maintenance jobs such as job orders and scheduling the day-wise maintenance to maintenance supervisors. Therefore, asset maintenance supervisors are accountable for job order management, resources management for performing the maintenance job, and documentation of jobs performed successfully. Asset maintenance managers must

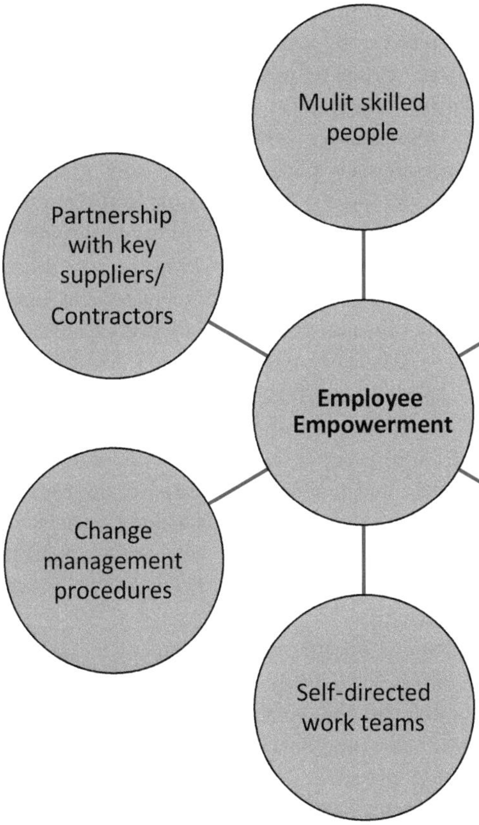

Fig. 2.8 Employee empowerment

facilitate the maintenance supervisors to achieve the plant's maintenance objectives at an organizational level.

The key contributing elements (Fig. 2.8) in employee empowerment are multiskilled people, operators-based maintenance, regular interaction with the staff, self-directed work teams, change management procedures, and partnership with key suppliers/contractors (Velmurugan, 2016).

Operator-based maintenance is the approach to empower the operators of equipment or plant to perform routine asset maintenance jobs. The operators must be trained on specific equipment/assets so that the routine maintenance jobs can be done timely, quickly, and properly. Developing multiskilled operators at operations-intensive organizations facilitates the improvement of the cost of asset maintenance activities and better operational efficiency.

In some of the huge plants such as petrochemicals, refinery, and production units, operator-based maintenance is not recommended due to the nature of operations and maintenance. In such plants, asset maintenance job orders are issued by operations engineer, and asset maintenance are carried out by maintenance engineer/

technicians. Otherwise, a sufficient workforce cannot be made available for performing operations and maintenance jobs separately in some small operations. In that case, **multiskilled people** are expected to complete the asset maintenance jobs and manage the plant/equipment's operations.

Regular interaction with maintenance staff is a key component in implementing the asset maintenance strategic plan. Therefore, such kind of interactions facilitates maintenance staff and operators to become **self-directed work teams**. The interactions can be planned interventions of regular weekly or fortnight meetings, providing technical and safety training to the frontline supervisors. It is evident from the research study that the maintenance managers in operations-intensive organizations are also using informal discussions to motivate and empower the maintenance staff and operators. There should be clear-cut standard operating procedures (SOP), emergency handling procedures, and regular briefing by supervisors/in-charges to empower self-directed work teams (Velmurugan, 2016).

A well-formulated asset maintenance policies and standard operating procedures (SOP) provide strength to operations, and maintenance teams in the organizations become self-directed in performing both routine maintenance and emergency jobs. A separate emergency preparedness procedure is available with most organizations in the industry for providing concrete steps to their employees working in operations and maintenance across the organization.

Change management procedure is another critical element to ensure proper documentation of minor modifications carried out by the asset maintenance team. There should be a systematic and detailed change management procedure to ensure the safe execution of modification. The modifications can be monitored centrally from main gas terminals, not by base terminals.

Partnership key suppliers/contractors must be facilitated with the definite and verified procedures like *delegation of authority* and *contracts and procurement procedures* in operations-intensive organizations. This kind of partnership provides timely help during emergencies in plant/facility and during shutdown maintenance. In case of nonavailability of spares from OEM, spares' requirements can also be handled with local vendors by suitable spare parts fabrication to avoid delay in maintenance and quick start-up of the plant.

Industry Quote Box 2.7: Employee Empowerment

- In our pipeline transmission operations, pipeline operations team goes to pipeline walk for observing the healthiness of the pipeline systems. If any minor repairs are to be undertaken in valves or any other components, the operators have the necessary skills to attend such repairs without calling maintenance technicians from the base station.
- Various safety permit systems like hot work permit, cold work permit, and work at height permit are followed in our facility. The operations team issues clearance for carrying out maintenance jobs through a permit system.

(continued)

Industry Quote Box 2.7 (continued)

This practice is to avoid any mishap in plant operations. For example, in case of electrical lockout permit, the electrical department requests that operation's shift in-charge approves the permit by obtaining the necessary clearance from the safety officer.

- While carrying out maintenance jobs, personal protective equipment (PPE) and non-sparking are mandatory in our operations plant premises.
- We maintain regular interaction with front supervisors through regular meetings every week. Daily meetings are conducted at control room in our process plant, wherein both operations and maintenance team meet and decide on day's maintenance plan.
- In our unit, the maintenance section in-charges review the status of maintenance schedule completion and resource requirements every week. Spares are maintained in shop floor to avoid delay in execution of maintenance jobs.
- Regular firefighting training is arranged by the safety department in our plant for all operators, maintenance technicians, and supervisors to empower them to handle emergencies. Similarly, security sensitization training is also carried out.
- Regular interaction among the maintenance personnel is ensured in our organization with various methods like daily progress reports being shared daily through the mail, on-the-job interaction with technicians, monthly review meetings, monthly safety meetings and safety campaigns, etc.
- Safety, security, technical, and firefighting training are conducted to our contract employees regularly since contract employees are the personnel directly involved in maintenance jobs in our process plant.
- Any modification in the plant systems is essentially pass-through change management system, written down procedure in our organization. The necessary approvals are to be obtained from the competent authority.

2.3.4 Information Technology

Information Technology developments such as Computerized Maintenance Management Systems are applied in asset maintenance management applications for effectively managing asset maintenance strategy formulation and implementation in operations-intensive organizations. CMMS application helps asset maintenance function across the organization to integrate with Enterprise Resource Planning system for improving asset maintenance performance by managing resources materials and workforce. CMMS facilitates the implementation of new production technologies that lead to global best asset maintenance practices.

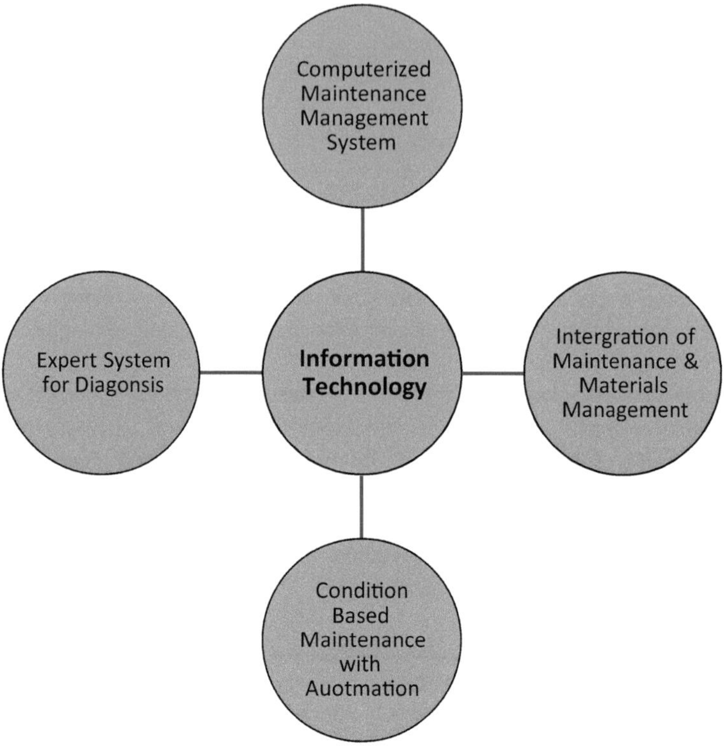

Fig. 2.9 Information Technology

The organizations become competitive among the industry peers and leverage e-maintenance tools and techniques to minimize the asset maintenance cost and maximize production. Work process improvements can be achieved by implementing these tools. Such improvements in asset maintenance management are maintenance job execution methods, the decision on maintenance job orders like a spare replacement and servicing the existing part, communication about the job after the execution, and monitoring the asset maintenance activities within the plant/ facility. These job improvement processes are achieved using IT infrastructure. Majorly, IT systems help achieve higher asset maintenance productivity, equipment/asset availability, and maintenance team effectiveness by impacting asset maintenance decisions to optimize maintenance cost for ensuring asset/equipment reliability.

Further, there are improved communications among the maintenance teams, OEMs, external subject matter experts with asset failure resolving support from the vendors and service providers. It is achieved through advanced information technology and communication technology even though the concerned stake holders are separated due to distances across the countries globally.

The key elements (Fig. 2.9) related to Information Technology with the specific reference to asset maintenance are the availability of the Computerized Maintenance Management System (CMMS), integration of maintenance and materials management, Condition-Based Maintenance (CBM) with automation, and expert systems for diagnosis (Velmurugan, 2016).

Availability of a **Computerized Maintenance Management System** captures every maintenance detail in a secure database from various units and master control rooms. Information can be accessed easily by multiple concurrent users from any computer or mobile device. It is used to prevent unplanned downtime and other threats from achieving optimal production. Processes and tasks can be streamlined. Access to historical data and important documents, as well as real-time updates on work order repairs, is available in the CMMS. It provides users the ability to have everything in one centralized, easy-to-access, and secure location. Such a system's benefits are job order management, preventive maintenance planning, asset and equipment management, safety monitoring, manage and control costs, labor management, reports, dashboards, mobile access, and regulatory compliance.

Integration of maintenance and materials management is another key element that requires asset management strategy implementation. The materials management systems need to integrate with the computerized maintenance management for inventory and spare parts management, purchase or service order management, material inspection, order closing, vendor/contractor management, and emergency handling systems.

Condition-Based Maintenance with automation is an innovative solution for asset maintenance function that uses technological solutions to capture real-time data of an asset/equipment to prioritize the maintenance decisions and optimize the asset maintenance resources. The technological solutions are Artificial Intelligence, machine learning, and blockchain on the IT platform Industrial Internet of Things (IIoT) that assist the asset maintenance team in making a timely decision with the available information. The various benefits of CBM with automation are asset/equipment failure early detection, asset maintenance activities optimization, remaining life assessment of an asset/equipment, asset utilization maximization, and a data-driven decision in asset maintenance function. The operations-intensive organizations adopt the Condition-Based Maintenance techniques benefit in numerous ways mentioned above and can move forward toward operational and maintenance excellence.

Industry Quote Box 2.8: Information Technology

- In our plant, most of the field instruments are converted to smart-based systems. An example is process transmitters like pressure transmitters, flow transmitters, and control valve positioners that are now intelligent instruments. All these instruments are connected to asset maintenance systems for capturing vital information from these sensors and do troubleshooting.

(continued)

Industry Quote Box 2.8 (continued)

- We use Supervisory Control and Data Acquisition (SCADA) system for operations of pipeline transmission systems and monitoring the critical parameters. In case of any alarm conditions, suitable actions are initiated from the control room.
- To leverage technology in pipeline systems, monitoring through satellites, Geographic Information System (GIS), and drones is implemented for monitoring healthiness of pipelines.
- SAP plant maintenance module is used for the whole bunch of asset maintenance activities in our process facility.
- In case of rotating equipment in our plant, vibration monitoring systems are integrated into asset maintenance systems for improving quick decisions and coordination with inspection plant.
- We are integrating our instrumentation control systems with the Industrial Internet of Things, even though the same is in the preliminary stage, and also, we are in communication with OEM of one of the critical equipment in our gas processing plant for implementing digital twin technology.

2.4 Asset Maintenance: Assessment Phase

In the assessment phase of asset maintenance, performance measurement of asset maintenance is a crucial factor, and the feedback from performance assessment can be used to improve the future asset maintenance strategy.

2.4.1 Maintenance Performance

The development of key performance indicators (KPI) for maintenance performance is an essential aspect of the performance process. It should be driven from the organization's corporate level to the plant level to ensure that asset maintenance activities are integrated with corporate objectives. The KPIs should capture the asset/equipment failure reasons and their occurrence and present them practically for understanding by asset maintenance managers.

Historical data of root cause failures of an asset/equipment over the various asset life cycle stages helps identify the asset maintenance's reliability improvement areas. These data must be the outcomes of maintenance performance measurements used by asset maintenance managers to improve the asset maintenance performance.

The KPIs based on real-time datasets indicate the business quality system, asset life cycle management, and maintenance practice benchmarking. Benchmarking must be the basis of establishing reasonable asset maintenance performance

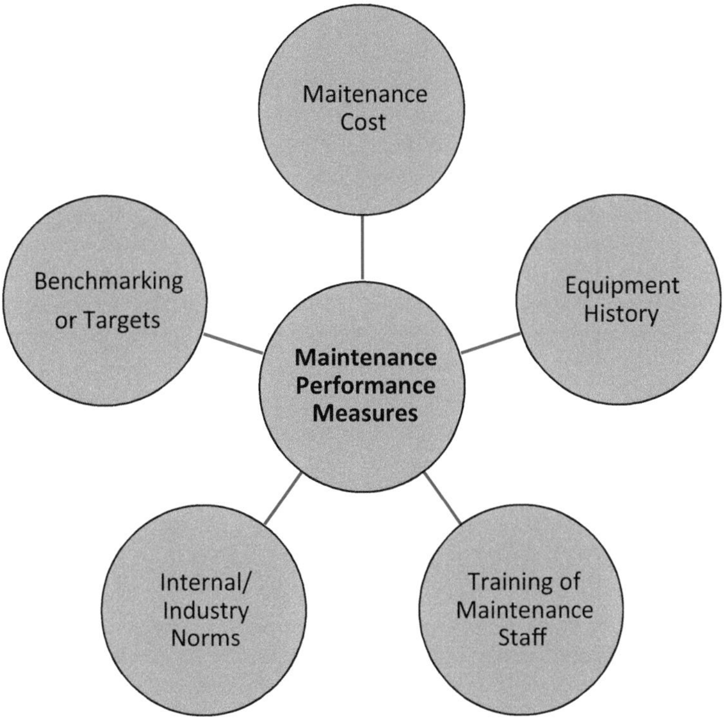

Fig. 2.10 Maintenance performance measures

objectives based on the global/national best practices of the industry. The key performance indicators and benchmarking parameters drive toward zero failure of asset/equipment, which embraces the best asset life cycle practices among operations-intensive organizations to achieve maintenance excellence.

The key elements (Fig. 2.10) to evaluate maintenance performance (MP) are maintenance cost, equipment history, training of maintenance staff, internal/industry norms, and benchmarking measures and targets. These are useful in formulating and selecting maintenance strategies based on maintenance performance measures (Velmurugan, 2016).

Maintenance performance measures show asset maintenance benchmarks or target completion status. It is important to understand how good maintenance jobs are performed and capable of minimizing equipment failures to improve equipment availability. These performance measures (Fig. 2.10) are statistics that assess the critical success factors which influence the anticipated result. Usually, lagging and leading indicators are used to understand operational assets' performance using the best maintenance strategies.

Maintenance costs impact the overall cost of the plant operations. The maintenance cost majorly includes workforce and material costs. Therefore, deployment of the optimum workforce in maintenance jobs shall be a key performance indicator of maintenance.

Further, the maintenance budget and actual expenditure of historical maintenance data (lagging indicator) are important while planning budget and spending on labor, material, stores, and department overheads. In general, organizations do not follow the cost spent per AM strategy. Therefore, maintenance cost expenditure per AM strategy is a leading indicator.

Training of maintenance staff is another essential parameter in maintenance performance measures. The maintenance staff is usually being trained from OEM of critical equipment. Even after training on critical equipment, critical equipment installations use expertise from OEM vendors. It is one of the key indicators to measure the training cost vs. OEM expertise cost (lagging indicator) to assess the training effectiveness as part of the maintenance cost. More efforts are required to develop in-house expertise by providing maintenance engineers an opportunity to build their expertise.

Equipment history including breakdown/shutdown records, trip analysis report (TAR) analysis on an annual/biannual basis, and other analysis reports is used as a lagging indicator to understand the equipment/plant availability and reliability. This indicator assesses the standby equipment's readiness (should be in hot standby, i.e., the equipment can be used in case of failure if the equipment is in-line) in manufacturing/production operations.

Internal/industry norms are used to assess the performance of AM in the way of conducting internal and external (third-party) audits based on internal and industry benchmarks. This parameter is used as a leading indicator. The internal norms are maintenance policy, safety policy, quality, and environment policy. Regular internal audits must be conducted on maintenance practices followed in organizations. The findings of the internal audits are useful to reduce operational risks and increase equipment availability for operations. Industry norms include national/international standards, standard operating manuals provided by OEMs, and local authorities' statutory guidelines. The external/third-party audits based on such industry norms offer insights about the maintenance standard and quality at an organizational level.

Benchmarking measures/targets are a parameter that needs to be considered while evaluating the maintenance performance measures. Therefore, targets are set at an organizational level or local government authorities' level to assess government companies' maintenance performance. The operations-intensive organizations must align the targets based on the following maintenance scheduling and completion, maintenance job quality, critical equipment reliability and availability, mitigation of operational risks, and optimum use of maintenance resources.

Industry Quote Box 2.9: Maintenance Performance

- Equipment-wise breakdown history is maintenance by plant maintenance personnel as part of asset maintenance records. These records are used to evaluate the maintenance performance of that equipment.
- We generally follow targets like plant availability and equipment availability as key metrics. These targets are set based on the production targets of the plant operations. The production targets are set stringently in line with the business objectives. As part of the maintenance team, we strive to achieve these targets by adopting better asset maintenance practices in our organization.
- In the case of the pipeline transmission system, benchmarking is the enforcement of statutory requirements since we are achieving the 99.9% pipeline availability. Therefore, we are doing benchmark on performance parameters.
- Our maintenance engineers are trained in various critical equipment by OEM suppliers. Even after such training, we call OEM experts for overhauling the critical equipment. Such experts' hiring costs contribute to the higher asset maintenance cost in the organization.
- Industry standards applicable to petrochemical and process plants are followed in greater spirits to maintain performance excellence in asset maintenance of the plant.
- In our organization, maintenance performance targets diverge with corporate-level targets. These targets are systematically broken down to department-level targets on an annual basis.
- Various internal audits by our cross-functional teams and external audits by statutory agencies or subject matter experts are the essence of maintenance performance in our organization. The audit recommendations and their implementation facilitate best maintenance practices, improvisation in maintenance processes, and adopting the standard's guidelines in true sense.
- Maintenance cost is one of the maintenance performance indicators in our facility, and it is benchmarked as per industry norms.

Maintenance scheduling and completion are to complete the job orders as per plan and priority. Documentation after completion of the job is a must, and the records are to be stored in any Computerized Maintenance Management System (CMMS). The equipment history must be possible to retrieve from the system as and when required in case of deviation due to operational issues and explicit approval from the concerned authorities or plant heads as per the company's internal guidelines. Backlog job orders are also to be completed as and when the opportunity arises.

Maintenance job quality ensures no repetition of the failure, failure due to poor quality, test plans after completion of the maintenance work, execution of job as per OEM recommendations from maintenance manuals, and proper usage of spare parts.

Critical equipment availability and reliability are to be ensured while performing asset maintenance jobs. For example, the replacement of locally fabricated spare parts in place of spares supplied by OEM may affect the equipment's reliability. Similarly, providing temporary solutions to the failures instead of identifying the root cause and resolving the root cause problem is one of the significant concerns in equipment/assets since it may fail again.

Operational risk mitigation is an essential indicator of maintenance performance. In natural gas pipeline transmission, proper maintenance and testing of pressure safety valves installed in gas compressors ensure its safety. Near-miss incidents on operational risk are to be recorded and analyzed against maintenance performance.

Optimum use of asset maintenance resources with techniques such as workforce sharing among units, centralized spare procurement, standard testing and workshop facility, and reducing the OEM expert requirement strengthens the asset maintenance team performance in all the aspects of asset maintenance management.

2.4.2 Maintenance Challenges

Asset maintenance managers from operations-intensive organizations consider that the maintenance challenges they face must be analyzed, and the required corrective actions are to be incorporated while formulating an asset maintenance strategy for effective implementation.

Maintenance challenges are the significant concerns (Fig. 2.11) related to asset maintenance like technical issues, contract issues, stakeholder issues, internal issues, and unexpected issues the asset maintenance team faces on a day-to-day basis (Velmurugan, 2016).

Maintenance challenge issues and their solutions are not generally considered while formulating an asset maintenance strategy since the problems are not systematically captured. Most of these issues are resolved by the individual and asset maintenance or operations team at the plant level, and the solutions remain as tacit knowledge with them. Therefore, there is an immense requirement in operations-intensive organizations that such tacit knowledge that exists with the experienced asset maintenance personnel requires to be captured systematically and documented for utilization of knowledge as and when needed. Few examples of maintenance challenge issues are described below but are not limited to those only.

Technical issues are related to the aging of the equipment and their behavior in the plant/facility's local environment. Change in equipment performance depends on the country's environmental conditions, place, and location apart from the equipment specifications. Due care must be given while designing assets/equipment to resolve such issues.

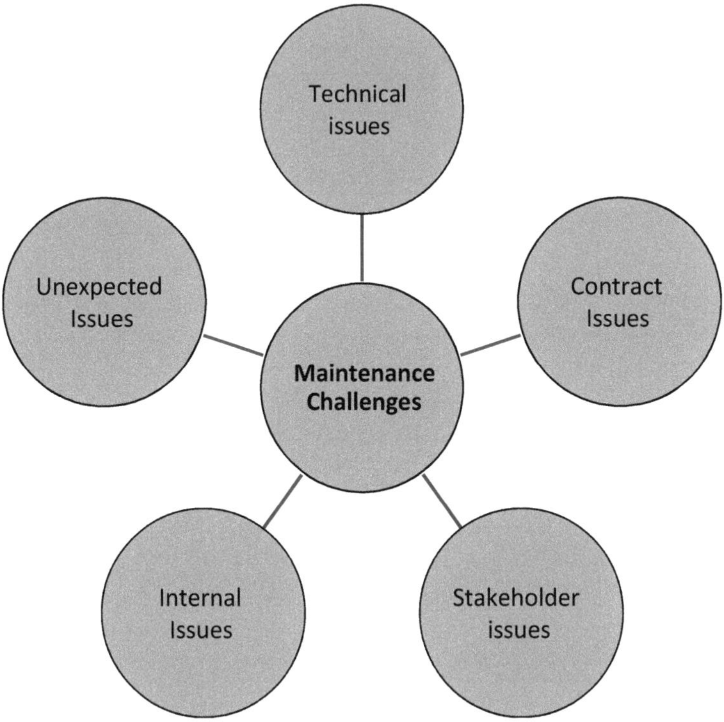

Fig. 2.11 Maintenance challenges

Industry Quote Box 2.10: Maintenance Challenges

- Aging of assets and critical equipment is challenging to the maintenance activities. The maintenance schedules are frequently requiring review due to this reason in our process plant.
- Our organizational philosophy is to deploy contract workforce in filed jobs. Therefore, the experience of such a workforce is critical in maintenance job quality. These contract workforces are deployed through workforce contracts. Once in 2 years, the workforce contracts are awarded freshly. In case of a new contracting company who is awarded with the contract, they may change the contract workforce deployed. The challenge to our maintenance team is to train the new workforce in our plant environment.
- We face few challenges in our organizations such as government policies, stringent procurement procedures, and outsourcing of manpower for various maintenance activities.
- In pipeline transmission systems, external stakeholders like farmers, approval from local government authorities, upstream suppliers,

(continued)

Industry Quote Box 2.10 (continued)

downstream consumers, etc. are critical in deciding any major maintenance job like leakages since all the stakeholders are to be convinced to carry out such job with necessary approvals from local authorities.

- In few instances, it is observed in our unit that maintenance job quality is affected due to inexperienced technicians who perform field jobs.
- All shutdown jobs are planned with the estimation of the timelines, even though maintenance personnel is kept under pressure to complete the maintenance jobs at the earliest citing the reasons of production targets or requirement of external stakeholders.
- Sometimes, maintenance personnel are required to take a decision on bypassing the system temporarily to avoid stoppage of the operations since rectification required shutdown of the operations. Much perseverance is needed to convince the concerned stakeholders in taking necessary approvals to bypass the system.
- In our organization, we have an emergency preparedness plan for a natural disaster like flood, earthquakes, and sabotage by terrorists. They are much critical and challenge to maintenance personnel.

Contract issues such as contract employee deployment vary whenever the contract changes from one vendor to another. The new contractor may deploy a new set of the workforce. It is a significant challenge for the asset maintenance managers to train such a contract workforce for plant-specific asset maintenance jobs. The suitable contract obligations are to be incorporated while placing the tender for the contract.

Stakeholder issues are from various stakeholders such as government, customers, suppliers, vendors, and contractors. Government policies and procurement guidelines are also challenging to asset maintenance managers to work within the framework's boundary. For example, oil and gas pipeline maintenance depends on the major stakeholder such as farmers and people occupying the land where the pipelines are laid.

Internal issues may be restrictions on procurement and resource availability. U**nexpected issues** such as terrorism, natural disasters, sabotage by criminals, and cyber threats are the major concerns for maintenance challenges. Suitable checks and balances and emergency procedures must be built into the asset maintenance strategic plan to resolve the internal and unexpected issues.

This chapter discussed deep insights on asset maintenance contributing factors based on rich theory captured from asset maintenance personnel working with operations-intensive organizations. Such deep understanding would help maintenance practitioners develop an asset maintenance strategic plan and adopt the organizations' best maintenance strategies and practices. Chapter 3 discusses asset maintenance strategies and practices in the industry.

Key Takeaways

- *Root cause analysis (RCA)* is an important technique that uses various tools and models to evaluate equipment/asset failure. This technique carries out complete analysis to dig out the root cause of the failure. The identified root cause of nonconformance in the system is removed by improving the processes.
- *Operator-based maintenance* is the approach to empower the operators of equipment or plant to perform routine asset maintenance jobs. The operators must be trained on specific equipment/assets so that the regular maintenance jobs can be done timely, quickly, and properly.
- *Change management procedure* is a crucial element to ensure proper documentation of minor or major modifications carried out by the asset maintenance team. There should be a detailed and verified change management procedure to ensure the safe execution of modification.
- The amount of backlog work orders is one of the important maintenance metrics. This metric is based on work type and coverage, schedule completion time, compliance, and backlog work hours.
- A maintenance policy is drawn from the maintenance mission and vision, a driving force to asset maintenance management in operations-intensive organizations.
- The aspects of asset maintenance performance cost of procurement and long-term maintenance plan could be improved to a greater extent by adopting a *centralized procurement method* in operations-intensive organizations where multiple units are geographically located across the country. This method could leverage by grouping the spares on types, criticality, consumption pattern, and lead time of procurement.
- The categorization of spares is critical spares and normal spares. Critical spares are spares of critical equipment/asset, spares with high procurement lead time, and high-value spares. Normal spares are consumables, spares with low procurement lead time, and low-value spares.
- The critical success factors of asset maintenance performance are the optimum deployment of contract workforce and expert workforce from Original Equipment Manufacturers.
- OEM supplies with standard equipment maintenance recommendations for carrying out periodic maintenance to the equipment in healthy condition. The recommendations are available in O and M manuals provided by the OEM.
- Predictive maintenance in operations-intensive organizations faces a paradigm shift due to the availability of intelligent field devices/sensors and the technology for remote connection of assets through a computerized network and the Industrial Internet of Things.
- Equipment criticality depends on how the equipment's failure impacts the organization's business, which is the basis of the reliability analysis of asset maintenance management. Therefore, identification equipment criticality must be carried out in operations-intensive organizations.
- Asset maintenance managers from operations-intensive organizations insist that the maintenance challenges they face must be analyzed and the required

corrective measures must be incorporated while developing an asset maintenance strategic plan.

SNAPSHOT: ASSET MAINTENANCE FACTORS

Asset Maintenance Contributing Factors

(1) Maintenance tactics; (2) reliability analysis; (3) maintenance performance; (4) materials management; (5) planning and scheduling; (6) Information Technology; (7) organization/HR resources; (8) maintenance policy/budget; (9) maintenance challenges; (10) employee empowerment.

Asset Maintenance Contributing Sub-factors

1. (a) Preventive maintenance; (b) predictive maintenance; (c) OEM recommendations; (d) Reliability-Centered Maintenance; (e) shutdown maintenance; (f) Total Productive Maintenance.
2. (a) Equipment history documentation; (b) critical equipment identification; (c) root cause analysis; (d) mean time between failures; (e) maintenance value risk study; (f) asset reliability statistics.
3. (a) Maintenance cost; (b) equipment history; (c) training man-hours; (d) internal/industry norms; (e) benchmarking/targets.
4. (a) Stores management; (b) purchase process; (c) stock management and materials scheduling; (d) optimization and rationalization of material process.
5. (a) Equipment history; (b) work order with priority; (c) preventive maintenance schedules; (d) weekly work schedules; (e) long-term/shutdown schedules; (f) measurement work backlog.

(continued)

6. (a) Computerized Maintenance Management System; (b) integrations of maintenance and materials management; (c) Condition-Based Maintenance with automation; (d) expert systems for diagnosis.
7. (a) Maintenance staff level; (b) maintenance organization; (c) responsibility of first-line supervisors; (d) technical training; (e) apprenticeship program; (f) deployment of contractors.
8. (a) Maintenance mission and objectives; (b) integration of maintenance plan and business plan; (c) maintenance policies; (d) maintenance budget.
9. (a) Technical issues; (b) contract issues; (c) stake holder issues; (d) internal issues; (e) unexpected issues.
10. (a) Multiskilled people; (b) operator-based maintenance; (c) regular interaction with maintenance staff; (d) self-directed work teams; (e) change management procedures; (f) partnership with key suppliers/contractors.

References

Mobley, R. K. (2008). *Maintenance engineering handbook*. McGraw Hill.

Velmurugan, R. S. (2016). *Doctoral Thesis. Maintenance strategy and practices at a large gas utility company in india—A case study approach*. University of Petroleum and Energy Studies.

Chapter 3
Maintenance Strategy, Practices and Spare Parts Management

Abstract This chapter deals with the selection and formulation of asset maintenance strategies for operations-intensive organizations. Many of the organizations' choice of a maintenance strategy are selecting suitable maintenance tactics/methods such as corrective maintenance, preventive maintenance, predictive maintenance, reliability-based maintenance, and Total Productive Maintenance. There is a need to view maintenance strategy selection in a holistic manner considering all the processes related to asset maintenance strategy and practices.

The key elements of asset maintenance strategy selection are maintenance personnel experience, Original Equipment Manufacturer (OEM) recommendations, maintenance policy guidelines, asset criticality and availability, and process severity and complexity. These are the major deciding elements in selecting a particular maintenance strategy for a specific equipment/process. The insights provided in this chapter are based on the knowledge captured from industry experts and professionals of operations-intensive organizations through semi-structured interviews.

The Maintenance Decision Support System is beneficial in selecting a maintenance strategy for multiple equipment systems discussed in this chapter. The interlinking roles of asset maintenance contributing factors with asset maintenance strategy and the importance of alignment of maintenance strategy with business and operation strategies are also deliberated.

The organizations must follow best asset maintenance practices while implementing the chosen maintenance strategy. The mission for best maintenance practices leads to contemporary asset maintenance management, which captures technical and operational datasets from the ecosystem and manufacturers' data to effectively implement the selected maintenance strategy considering the condition of the asset/equipment and the system's overall life cycle. A typical contemporary maintenance program attempts to optimize old-style maintenance practices and incorporate condition assessment of the asset, historical information, and other best practices. It is recommended to follow a nine-step process of the organization's asset maintenance practices to implement the selected maintenance strategy effectively. This chapter also describes the applications of OR and MS concepts in spare parts management and maintenance decision-making.

R. S. Velmurugan, T. Dhingra, *Asset Maintenance Management in Industry*,
International Series in Operations Research & Management Science 310,
https://doi.org/10.1007/978-3-030-74154-9_3

3.1 Asset Maintenance Strategy Frameworks: An Overview

An asset maintenance (AM) strategy is constructed on asset maintenance aims and policies of the organization. The asset maintenance goals are the basis for establishing plant-level targets to implement the maintenance strategy. Asset maintenance strategy includes the following key characteristics:

- Customer requirements are the basis for various business targets like production performance, product quality, and logistics planning.
- To execute asset maintenance strategic objectives, making available all the necessary resources must be at an organizational level.
- The asset maintenance team's performance on their functional capabilities is supervised, controlled, and evaluated for the job delivery.
- AM strategy must be agile enough to adopt the changing environment in the industry.

The AM strategy framework must guide the development of an asset maintenance strategic plan to effectively implement asset maintenance activities that provide substitutions, direction, and flexibility enough to the industry's varying atmosphere. This strategic plan must be developed considering the best and benchmarked practices available in the industry and suited to the organization's asset maintenance objectives. The successful implementation of the asset maintenance activities depends on the well-designed asset maintenance strategic plan. The strategic plan includes various aspects of an organization like its mission and vision, asset maintenance objectives and targets, and internal policies.

The frameworks related to the area of plant asset maintenance available in the literature, like journals/books found useful in asset maintenance of operations-intensive organizations, are described in this chapter. It is essential to comprehend the critical elements of the asset maintenance strategy described in these frameworks. The elements discussed in the frameworks can be used while selecting and formulating maintenance strategies at an organizational level.

The asset maintenance-related frameworks/models are reviewed critically in terms of formulation of asset maintenance strategic plan and its outcomes, choice of asset maintenance strategies formulated, and the execution of chosen maintenance strategies and maintenance performance. The description and critical concepts of asset maintenance strategy frameworks/models and their inferences related to operations-intensive organization are discussed below.

A framework on asset maintenance strategy formulation and review (McAllister et al., 1999) explains that the asset maintenance function must be considered a collaborator within the business. Such an approach leads to achieving operational profits in the organization by aligning the corporate-level objectives with the asset maintenance function's objectives. The successful implementation of asset maintenance strategy depends on various components such as organization targets, operation and maintenance policies, best maintenance practices, the scope of the asset maintenance strategic plan, and benchmarking. Asset maintenance strategy's key systematic aspects are supported in terms of resources to asset maintenance, audits on strategy implementation, and organizational aspects like employee development, skill training, and employee empowerment. The framework covers defining business objectives, allocating adequate resources to perform asset maintenance, monitoring asset maintenance strategic plan outcomes, and flexibility to adapt to changes in the industry quickly.

Business-centered model (Kelly, 2006) describes that the organization's corporate aims are developed; thus, the asset maintenance objectives are built. These objectives facilitate the formulation of an asset maintenance strategy. Operations of the plant/facility are based on market conditions, i.e., production targets are decided based on market requirements. Therefore, the asset maintenance objectives are to be derived from operations and corporate objectives. The BCM model suggests various factors for implementing asset maintenance strategies such as asset maintenance plan for entire asset life cycle duration, resource planning for executing asset maintenance plan, asset maintenance organizational structure and monitoring, and control systems. The BCM model is also validating the requirement of aligning corporate, production, and asset maintenance objectives systematically.

Another framework is on the **schematic of asset maintenance processes while formulating the maintenance strategy and its structure** (Salonen, 2011). The asset maintenance strategy alignment, developing maintenance performance indicators, and timely completion of asset maintenance strategic plan with the targets are the key elements described in this framework. This framework also explains the importance of asset maintenance strategy's alignment with corporate strategy and KPIs of asset maintenance.

A framework on **asset maintenance management** (Pintelon & Parodi-Herz, 2008) describes various aspects of asset maintenance strategies such as system design, planning and control, and administrative tool kits. The system design includes asset maintenance policies and manufacturing control systems. The planning and control of asset maintenance comprise the asset maintenance objectives, resources to achieve these objectives, and performance measurement and reporting. The asset maintenance administrative tool kits are IT infrastructure and asset failure modeling techniques. This framework covers the complete asset maintenance decision-making process from strategic planning to operational planning. It also emphasizes the criticality of optimization and combining these planning processes at an organizational level.

A strategic maintenance performance management framework (Eti et al., 2006) describes aligning the business strategy with asset maintenance strategy at an organizational level. The framework emphasizes asset maintenance implementation plan, measurement of asset maintenance performance, and feedback to business strategy. The critical elements related to the asset maintenance strategy explained in this framework are asset maintenance organization structure, efficacy measurement of asset maintenance, control over maintenance jobs, logistics support, computerized asset maintenance management systems, personnel records, and asset maintenance engineering-related tasks.

The performance measurement framework for asset maintenance function (Muchiri et al., 2011) highlights the activities regarding the alignment of the corporate strategy to be taken care of while developing the asset maintenance strategy. This framework suggested leading asset maintenance performance indicators such as planning and scheduling, job execution, job identification, and lagging asset maintenance performance indicators such as benchmarks, performance targets, and asset maintenance costs. Therefore, this framework focuses on asset maintenance performance with a few concepts of asset maintenance strategy formulation.

A framework for asset maintenance strategic planning (Umar, 2011) explains nine functions of asset maintenance strategy such as major stakeholders' recognition, mission statement formulation, strategic objectives establishment, baseline condition assessment, strategic issues identification, available strategic options, selection of maintenance strategy, performance measures development, and strategy implementation planning. All these functions cover the asset maintenance strategy cycle from asset maintenance strategy development to execution.

A common thread in the above-discussed asset maintenance management frameworks/models is aligning the organization's corporate-level objectives with plant-level objectives for profitable business operations. All these frameworks are lacking a holistic approach of covering all the asset maintenance strategy processes such as the development of asset maintenance strategy, identification of suitable asset maintenance strategy for a specific asset/equipment based on its type and requirements, implementing the chosen asset maintenance strategy, performance measurement of strategy implementation, feedback improvements back into the initial stage of asset maintenance formulation, and finally company-wide integration of the asset maintenance-related activities.

From the above discussion on various asset maintenance strategy-related frameworks, it is inferred that each of the above frameworks describes a few specific processes of maintenance strategy but not in a combined manner integrating all the maintenance strategy-related processes. Based on these inferences, it is evident that a holistic model that combines all the asset maintenance processes from asset maintenance strategic plan development to implementation is beneficial to asset maintenance managers/engineers in operations-intensive organizations. AM strategy constituents such as maintenance strategy selection, formulation of maintenance strategy, and implementation of such maintenance strategy and asset maintenance practices are discussed in this chapter.

3.2 Asset Maintenance Strategy Selection

In general, many of the organizations' selection of maintenance strategy are a selection of suitable maintenance tactics/methods such as corrective maintenance, preventive maintenance, predictive maintenance, reliability-based maintenance, and Total Productive Maintenance. There is a need to view maintenance strategy selection in a holistic manner considering all the processes related to asset maintenance strategy and practices.

3.2.1 AM Strategy Selection Process

In operations-intensive organizations, the significant elements of asset maintenance strategy selection are maintenance personnel experience, Original Equipment Manufacturer (OEM) recommendations, maintenance policy guidelines, asset criticality and availability, and process severity and complexity (Velmurugan, 2016). These are the deciding elements (Fig. 3.1) in selecting a particular maintenance strategy for a specific equipment/process.

Maintenance personnel experience plays a critical role in deciding the maintenance selection process for a specific process/plant/equipment. In most situations, even though well-defined maintenance methods are available, selecting a particular maintenance method for a specific process is a crucial decision for a maintenance manager. This decision differs on various issues such as the organization's culture, size of the organization, maintenance objectives, behavior of the specific process,

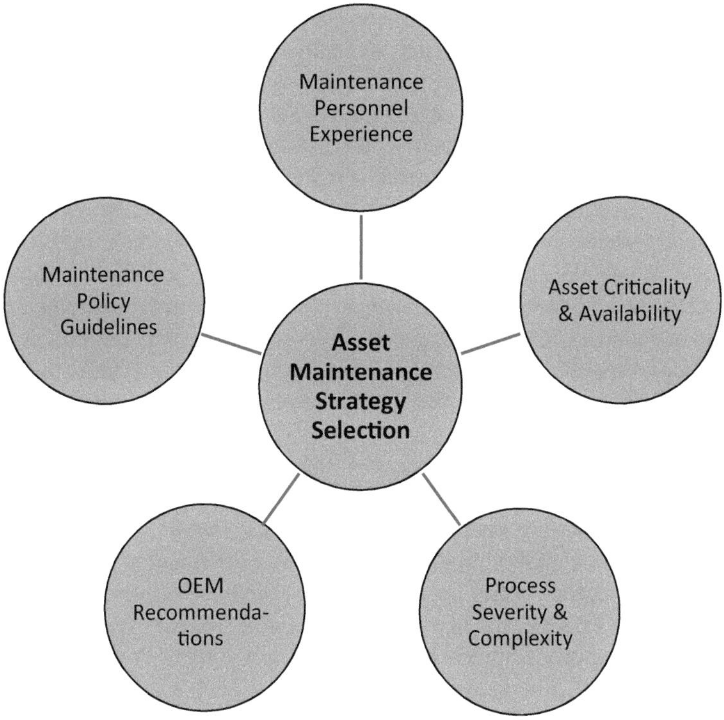

Fig. 3.1 Asset maintenance strategy selection process

and equipment in the country/region. Therefore, tacit knowledge available with maintenance personnel is crucial to decide the maintenance strategy at an organizational level.

Asset criticality and availability are other critical elements in the maintenance strategy selection process. The most of the organizations are carrying out critical assets identification. However, they fail to completely comprehend criticality ranking parameters. Criticality is decided based on plant objectives, operational targets, quality of the product, and production cost. In any operations-intensive organization, the major thrust is given to operations. Therefore, a specific asset-wise characteristic is overlooked initially. Because of this constraint, asset criticality must be assessed covering the following: historical records of preventive maintenance, breakdown maintenance of the specific asset, mean time between failures (MTBF), failure probability index, spare procurement lead time, asset replacement value, OEM recommendations, safety and atmosphere impact, skill to separate single-point letdowns, and stakeholder's influence. The organizations need to assign weighted average factors to all these factors according to their requirement and maintenance strategy selection objectives set at an organizational level.

Process severity and complexity deal majorly with a wide variety of risk and compliance issues in operations-intensive organizations. These issues can complicate processes, slow down process efficiency, and finally reduce profitability. It is essential to ensure built-in process safeguards without neglecting maintenance. According to stringent operational parameters, the assets related to operations must perform at their peak efficiency by mitigating plant risks like valve leaks, pressure and temperature parameter limits, and wastages. The other factors related to process severity and complexity are health and safety, environmental impact, local government laws and rules, emergency and disasters, and technological upgrading.

Original Equipment Manufacturer (OEM) recommendations must be considered while selecting an asset maintenance strategy of equipment. OEM suggests specific maintenance requirements for each equipment in their operations and maintenance manual. These recommendations include preventive maintenance after specific running hours of equipment, replacement of materials/spares periodically, and standard maintenance practices. In comparison, the selection of maintenance strategy for particular equipment is based on the recommendations by OEM. The recommendations are provided in the operations and maintenance manual supplied by OEM.

Maintenance policy guidelines are essential in choosing a maintenance strategy for plant asset maintenance. The guidelines include equipment in a ready and reliable condition to meet varying operational requirements. Further, equipment function must always ensure a quality product and no sudden or frequent breakdown of the equipment. Maintenance policy ensures no major breakdowns, optimum inventory, asset replacement strategy, and least maintenance cost. In case organizations cannot fulfill the maintenance guidelines, it negatively impacts the following: full capacity utilization, increased production cost, increased maintenance cost due to spares consumption, reduced product quality, and the worst impact on the safety of the plant operators.

Industry Quote Box 3.1: Maintenance Strategy Selection

- Maintenance strategy selection in our plant is carried out considering the process severity and criticality of the equipment. We use standby equipment strategy in critical operations.
- In our company, maintenance policy provides broader guidelines for selection of maintenance strategies. The maintenance strategy selection depends on process complexity, equipment criticality, availability of equipment as well as plant, and cost of maintenance.
- Preventive maintenance and predictive maintenance are two major maintenance strategies practiced in our facility. The other factors that help decide the maintenance strategy are the experience of the original equipment manufacturer's maintenance personnel and recommendations.
- In case of operations-intensive organization like petrochemicals, the predictive maintenance is used for long-term or shutdown maintenance planning. Preventive maintenance is used for short-term maintenance planning. Asset integrity management concepts are being implemented based on industry standard.
- We have multiple processing plants across the country. The maintenance strategy adopted based on various root cause failure studies in one plant is shared among all the processing plants so that they can quickly adapt such practices to avoid similar kind of failures in their plant.
- Maintenance strategy selection is completely guided by industry standards and company's internal norms in our company.

The following factors play a role: industry-related standards, government policies, internal norms such as sharing maintenance problems among work centers at an organizational level, geographical zones, countries, and CMMS usage level. These factors help the organization implement a uniform maintenance strategy selection process across its work centers located within a country or across countries.

3.2.2 Maintenance Decision Support System

The suitable asset maintenance strategy from the chosen strategy for equipment or specific to the process/operation can be selected using the Decision Support System (DSS) models. Such models can be optimized for achieving asset maintenance excellence with reasonable maintenance costs. There are specifically developed Plant-Level Maintenance Decision Support Systems (PMDSS) that are available. These systems consider both long-term and short-term asset maintenance objectives aiming at zero downtime, maximizing the equipment availability with optimum resource allocation while achieving the equipment's maximum capacity (Li et al., 2009).

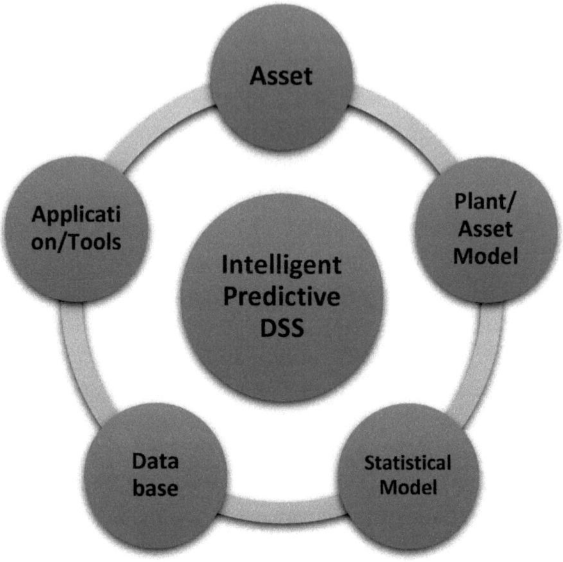

Fig. 3.2 Intelligent predictive Decision Support System components

Further, in the Operations Research subject, asset maintenance selection is contemplated as a mathematical model, and it can be solved to identify the suitable asset maintenance strategy for asset maintenance optimization in the industry (Velmurugan & Dhingra, 2015).

The asset maintenance managers must choose the best asset maintenance strategy for particular equipment or a system out of a set of maintenance strategy alternatives. It is a tricky job to identify the best set of asset maintenance strategies, i.e., a suitable combination of preventive, predictive, intelligent maintenance methods for different failure modes. The maintenance strategy formulation depends on various asset maintenance contributing factors discussed in Chap. 2. Decision Support Systems uses multiple statistical tools and techniques to select a suitable maintenance strategy through a systematic process.

The primary objectives to be considered in PMDSS are as follows:

- The system must consider the following asset/process parameters as asset availability and criticality: the life cycle of the asset, asset safety, and process severity.
- Another important objective is to optimize the asset maintenance cost by maximizing asset uptime, production capacity, product quality, and sustainability.

In the current technological scenario of the globe, there are intelligent Decision Support Systems (Fig. 3.2) that are available to predict the asset maintenance strategy in terms of different operational requirements of the equipment such as equipment being operated in normal process conditions, maximum capacity utilization, degrading of few equipment parameters, and reoccurrence of equipment failures. According to these conditions, asset/equipment maintenance action varies based on specific maintenance requirements, no particular maintenance action, and condition monitoring. Therefore, prediction models built into the PMDSS would take appropriate action from the library of the maintenance activities mapped in the system. Further, the Decision Support System must choose between planned maintenance action or predictive model (mathematical, neural network, and logical methods) based on maintenance action. The action depends on the case-by-case basis of the asset condition or requirement.

The above-discussed MDSS systems help optimize maintenance costs by suitably initiating the maintenance actions before the possible fatal breakdown of equipment/asset. The asset/equipment failures can be detected before such failures become critical before major plant/operations shutdown.

3.2.3 Specific Maintenance Strategy

The selection of specific maintenance strategies in operations-intensive organizations is based on the following significant factors (Fig. 3.3): Original Equipment Manufacturer (OEM) recommendations, industry standards, experience by the maintenance managers, and maintenance policies.

OEM recommendations focus on preventive, corrective, and overhaul maintenance programs for the asset manufactured by the concerned OEM, and all the recommendations are prescribed in their operations and maintenance manuals. Such OEM documentation helps build the asset-based maintenance program to maintain the asset's healthiness during its life cycle. The asset's reliability and safety are the significant risk mitigation activity of the specific asset/equipment maintenance strategy. The OEM recommendations are also contractual, statutory obligations between the asset owner and the OEM for performing Condition-Based Maintenance. Condition-Based Maintenance program consumes many resources such as spare parts, consumables, and other materials, which led to higher operation and maintenance costs. Therefore, it is crucial for OEM's recommended maintenance program to be supported by its business and operational purposes. The asset owner understands the OEM recommended asset maintenance program's effectiveness based on their experience and decides to revise it in consultation with OEM.

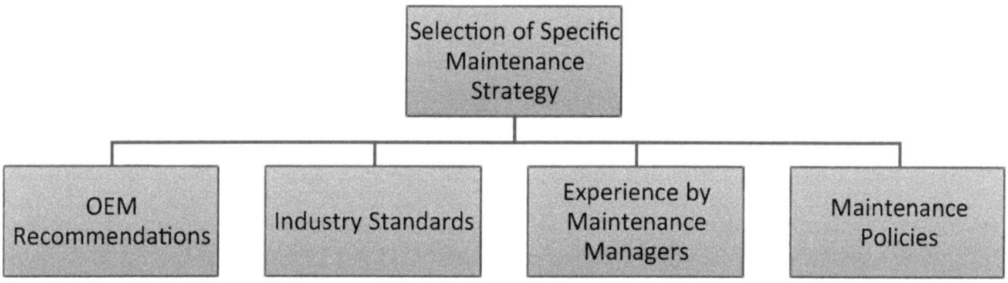

Fig. 3.3 Selection of specific asset maintenance strategy

Industry standards are not like OEM recommendations. The standards are not proprietary. It is common to all the assets/systems since the standards are tried-and-true, tested, and validated. Therefore, asset owners can incorporate the standards for their purposes in asset maintenance activities. Such standards have offered benefits such as interoperability across assets, easy-to-replicate asset maintenance procedures, best practices among all the assets across all geographical locations, and technological endurance.

The industry standards also include maintenance tasks such as cleaning, lubrication, spare replacement, repair strategy, and asset data history utilization. The standards also provide checklists for maintenance tasks, maintenance schedules, and integrated maintenance programs. The standards-based asset maintenance management provides systematic asset maintenance practices in a diverse landscape of different process plants, asset installations, equipment, and processes. There is a host of asset maintenance-related standards available. Few examples of standards developed by the institutions are the International Standards Organization (ISO), American National Standards Institute (ANSI), American Gas Association (AGA), American Petroleum Institute (API), European Standards (EN), etc. ISO 55000 Asset Management Standards is a relevant standard for asset maintenance management.

Maintenance policies in an organization are drawn from the OEM recommendations and industry standards to provide maintenance managers and engineers guidelines for designing the asset/equipment-wise, appropriate maintenance program. The organization-level maintenance policy ensures the following aspects: reliability of assets, the healthiness of equipment to maximize availability, stop unwanted and frequent breakdown of equipment, optimization of production capacity, and its quality and control on maintenance cost. A well-defined maintenance policy is crucial for operations-intensive organizations.

In addition to the above-described factors, the **maintenance manager's experience** also plays a critical role in selecting a specific maintenance strategy for a particular equipment/asset. There are many experiences on previous maintenance works about the asset, which is not documented and resides as tacit knowledge with the maintenance managers/engineers. Even though organizations deploy various systems to tap such knowledge through Computerized Maintenance Management System (CMMS) applications, SAP platform, and other asset maintenance systems, it is still challenging for them to tap such tacit knowledge. Therefore, while selecting an appropriate maintenance strategy for a particular equipment/asset, it is important to consider the concerned maintenance manager's view. Such an approach helps select a suitable maintenance strategy that leads to maintenance excellence at an organizational level.

3.3 Asset Maintenance Strategy Formulation

3.3.1 Role of AM Contributing Factors

Asset maintenance contributing factors are discussed in detail in the previous chapter (Chap. 2). All these factors are playing a critical role in formulating an asset maintenance strategy. In this section, the connections of their role are described further to develop a method in AM strategy formulation, and the outcome of this exercise is the AM strategic plan. The elements drawn from each AM contributing factor which must be given suitable importance while formulating organizational-level AM strategic plan are tabulated in Table 3.1.

3.3.2 Alignment with Business and Operation Strategies

The alignment of asset maintenance strategy regarding the complete asset life cycle with the operations is crucial for effective AM strategy implementation in operations-intensive organizations. Maintenance performance metrics discussed in Chap. 2, like maintenance cost, equipment history records, and asset utilization, are essential to understand equipment/asset healthiness in the organizations. The asset life cycle process needs to be understood in terms of these metrics.

The following are the significant objectives to be considered while aligning AM strategy with business and operational strategies:

- To understand the asset's current position in the asset life cycle and its expected end of life
- To evaluate the asset's current condition regarding the end of life of the asset and the availability of its complete history
- To measure the maintenance cost of a particular asset and its work output in terms of cost value
- To make available asset management database for monitoring the asset's healthiness throughout the asset life cycle

Typically, organizations are interested in measuring the sunk costs of assets deployed. Optimization of total costs over the entire asset life cycle stages is one of the methods to achieve an organization's business objective while maintaining the availability of the asset to operations. Another method is the formulation of a cost-effective AM strategic plan across all the asset life cycle stages. The maintenance cost is also optimized across all the stages of the life cycle. Therefore, the best approach is a suitable maintenance strategy that allocates maintenance cost-efficiently across the asset life cycle stages to optimize maintenance costs while performing operational objectives and achieving business objectives.

AM strategy is selected and formulated first and then implemented at an organizational level in operations-intensive organizations. It becomes essential to measure

Table 3.1 Role of AM contributing factors

AM contributing factor	Elements deciding AM strategy formulation	Explanation
Maintenance policy/budget	• Equipment readiness and reliability • Equipment healthiness • No major breakdowns • Maintenance cost • Maintenance obsolescence • Maintenance audits • Maintenance revenue budget • Maintenance capital budget	In case the organization policy says after every 10 years, the equipment can be replaced under the element of maintenance obsolescence. It is a guiding factor for the maintenance manager to formulate a maintenance strategy. Accordingly, the amount for the replacement of equipment in the capital budget needs to be provisioned. Similarly, maintenance cost must be less than 1% of gross profit or net profit. Therefore, a maintenance revenue budget is to be provisioned and an equipment maintenance strategy to be formulated
Materials management	• Emergency purchase procedure • Spare inventory and availability • Critical/insurance spares • Inventory and analysis control	Materials management plays a vital role in AM strategy formulation. For example, the availability of spares is essential in formulating preventive/predictive maintenance approaches. Critical spare like gas turbine rotor must be readily available in case of major equipment failure due to rotor damage. The unexpected equipment failures are managed through emergency purchases by maintenance practitioners
Organization/ human resources	• Training through OEMs on critical equipment • Knowledge management initiatives • Skilled workforce • Deployment of contract workforce	Proper training through OEMs on critical equipment reduces OEM expert dependency for a major overhaul of critical equipment. Knowledge management initiatives help an organization to tap tacit knowledge available with maintenance managers. The skilled workforce within the organization and deployment of contract workforce must be understood while formulating AM strategy, and workforce requirements must be incorporated suitably in the maintenance management program
Planning and scheduling	• Equipment history • Maintenance schedules • Shutdown maintenance • Short-term planning	The equipment history is an important feedback on maintenance carried out over the years. Therefore, the outcomes from such a history report are to be incorporated in AM

(continued)

Table 3.1 (continued)

AM contributing factor	Elements deciding AM strategy formulation	Explanation
		strategy's formulation. Accordingly, maintenance schedules are to be revised. The critical differentiator in maintenance strategy formulation is long-term shutdown or short-term maintenance planning
Maintenance tactics	• Preventive maintenance • Predictive maintenance • Reliability-Centered Maintenance • Total Productive Maintenance • Inspection-based maintenance • Asset-based maintenance	There are various maintenance tactics available for asset maintenance. One of the primary objectives of the asset maintenance strategy formulation is to identify suitable maintenance tactics or a combination of maintenance tactics for the particular asset/equipment/process. Accordingly, an implementation plan is developed
Reliability analysis	• Critical equipment identification • Root cause analysis • MTBF/MTTR	Critical equipment identification is the basis for specific maintenance strategy formulation for the asset undertaken for AM strategy formulation. The other two elements such as Root Cause Analysis (RCA) and MTBF/MTTR are helpful to revisit the AM strategy formulated, and if it requires revised AM strategy can be formulated
Information Technology	• CMMS S/W applications • Integration of maintenance and materials management modules in the S/W application	The Computerized Maintenance Management System software applications help formulate AM strategy systematically, and integration with materials inventory helps plan specific AM strategy
Maintenance performance measures	• Internal/industry norms • Benchmarking/targets	These two elements are guiding elements for AM strategy formulation. AM strategic plan is drawn based on internal norms such as maintenance policy and industry norms such as national/international standards
Maintenance challenges	• Technical, contract, stakeholder, and unexpected issues	While formulating AM strategic plan, care must be taken by maintenance managers to minimize maintenance issues. Suitable interventions must be incorporated in the plan based on their experience and equipment/asset history documentation

whether AM strategy is working optimally. Two measures to know the optimal implementation of AM strategy are efficiency, i.e., how well the maintenance team implement its intended strategy, and effectiveness, i.e., whether such implemented AM strategy yields desired outcomes or not. These measures can optimize the ROI on asset maintenance investment aligning with the business objectives of the organization. The efficiency improvement yields incremental profits, while effectiveness improvement drives strategic gains for the organization.

Organizations are measuring maintenance performance in isolation from the broader business context. If the asset/equipment is available for operations whenever required, then the AM strategy is perceived to be effective. However, it is important to understand the asset's value to measure the AM strategy plan's effectiveness regarding the maintenance cost associated with that asset from the corporate strategy perspective. The best practice is to optimize the maintenance cost and overall production cost to decide on the AM strategic plan.

The asset maintenance database is majorly involved in creating equipment information repositories and retrieval on a timely basis. This repository includes asset-related information such as asset details, equipment parts, components, specifications, datasheets, maintenance methods, maintenance schedules, standard operating procedures, and standard servicing procedures. Such a database helps to formulate an effective AM strategic plan and a well-understood plan, in turn, to align with the operational and business plans of the organization.

The above-explained aspects help operations-intensive organizations guarantee that AM strategy, and best maintenance practices apply reliably across all the organization's similar assets. This kind of approach enables the maintenance team to ensure similar assets are managed similarly, i.e., standardization of the AM strategy.

3.4 Asset Maintenance Practices

Asset maintenance practices are the best maintenance practices being followed by an organization while implementing the chosen maintenance strategy. The mission for best maintenance practices leads to contemporary asset maintenance management, which captures technical and operational datasets from the ecosystem and manufacturers' data to effectively implement the selected maintenance strategy considering the condition of the asset/equipment and the system's overall life cycle. A typical contemporary maintenance program attempts to optimize old-style maintenance practices and incorporate condition assessment of the asset, historical information, and other best practices.

The maintenance cost varies based on different maintenance strategies chosen for a specific asset/equipment/process and criticality. The optimization of maintenance costs is the trade-off between the direct cost of maintenance expenses and the benefits of failure cost avoidance. The cost avoidance measures include regulatory penalties in case of safety and environmental impacts, operational revenue losses due to asset unavailability, and societal implications of asset failures.

In the case of safety-critical equipment, the cost of significant accident hazards compensates the cost of industry best practices in preventive and predictive maintenance, and for this reason, it should not be accomplished with maintenance practices. However, for less critical assets/equipment, maintenance costs can become comparable or lower than repair/replacement costs. In many of the cases, the initial asset failure costs are not considered in the budget. Therefore, these expenses are required to be accommodated in other O and M expenses.

3.4.1 Best Maintenance Practice Steps

The prime aim of the best maintenance practices is to ensure safe operations and maintenance at an organizational level. The significant points to be considered in achieving the best maintenance practices are as follows:

- Deferred maintenance must be avoided so that asset/equipment is maintained healthy to mitigate any potential hazard due to deferred maintenance.
- Timely performed maintenance activities facilitate the life expectancy of asset/ equipment.
- Optimization of maintenance costs in line with the best maintenance practices is a must activity.
- Maintenance strategy and practices should rectify failures quickly rather than be proactive in their response and correct failure preconditions before becoming potential failures.
- The best maintenance practices should minimize callbacks and allow more time for planned maintenance.
- A healthy maintenance practice depends on the excellent maintenance crew and their responsibility, capability, and accomplishment.

The steps recommended for best maintenance practices are depicted in Fig. 3.4.

It is required to collect all the possible datasets regarding equipment downtime, MTBF, spares consumed, technician response time, technology utilization, and failure recovery time. Based on these, the average maintenance cost per hour for asset/equipment downtime can be calculated. This cost helps the maintenance manager to decide on the effectiveness of maintenance. Subsequently, the effect of average maintenance cost per hour on production cost can be calculated to identify the specific asset/plant/process's maintenance strategy's value.

Further, the timely availability of spares increases the response time of attending to the maintenance crew's equipment failure. A well-planned Computerized Maintenance Management System (CMMS) impacts job order management. It facilitates an increase in maintenance crew response time and reduces the MTTR and downtime of an asset. A well-structured preventive maintenance schedule covering all the critical equipment/assets in line with the OEM/process licensors' recommendations

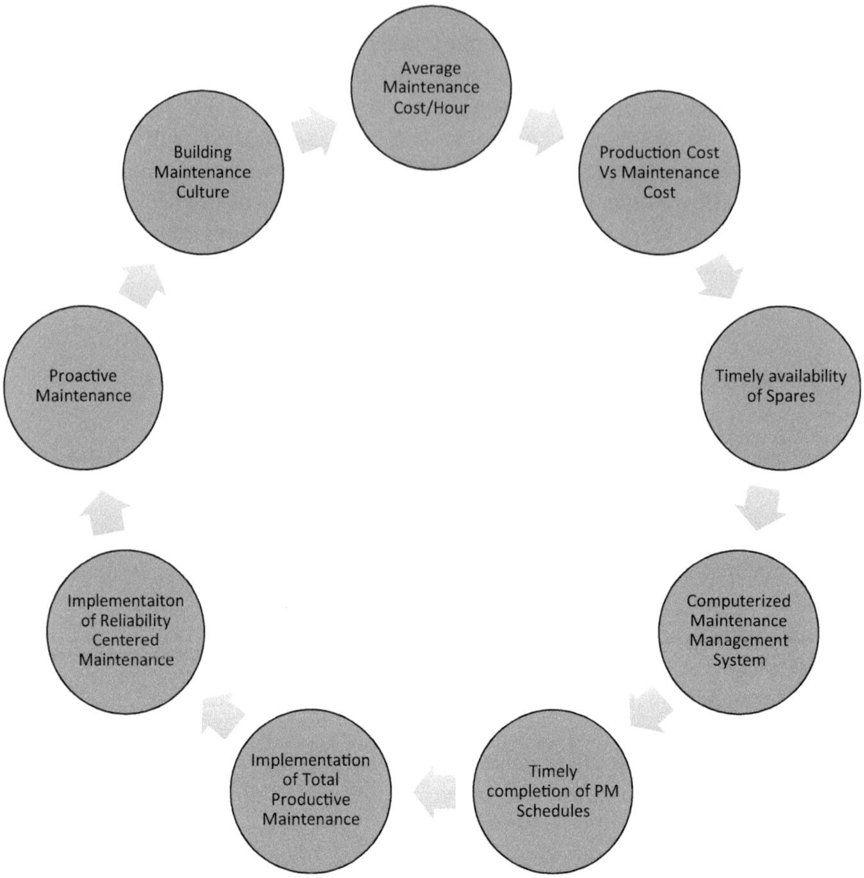

Fig. 3.4 Asset maintenance practice steps

and timely completion of such schedules without any deviation contributes to the best maintenance practice.

Best maintenance practices change the maintenance crews' mindset from reactive maintenance to proactive maintenance practices based on predictive tools. The Total Productive Maintenance approach takes forward maintenance practices to the next level by involving operators in "operator-based maintenance."

Further, Reliability-Centered Maintenance (RCM) helps to go to another level of best maintenance practice, i.e., no longer equipment brought down for planned preventive maintenance on a scheduled basis; instead, equipment run to the threshold of failure to assure increased production capacity with the help of prediction model.

Finally, *building a strong maintenance culture* is the need of the current volatile business scenario. The maintenance culture is built around proactive maintenance crew, e-maintenance practices, and improvements in technology. A framework for asset maintenance strategy and maintenance practices planning and execution for operations-intensive organizations is developed based on the above concepts on asset maintenance strategy formulation, selection of asset maintenance strategies, and best maintenance practices in the industry and described in Chap. 4. The framework suggested in the next chapter comprehensively covers all the aspects of asset maintenance strategy and practices and provides a holistic approach for maintenance practitioners to develop their customized AM strategic plan for their organizations in the industry.

3.5 Spare Parts Management

OR and MS concepts related to asset maintenance management are discussed in Chap. 1. The concepts of MRO supply chain management and inventory and asset tracking are the enablers to spare parts management. The experiences shared by asset maintenance managers in respect of these concepts during the research study conducted among operations-intensive organizations are presented below.

Optimizing the maintenance, repair, and operations (MRO) supply chain in the operations-intensive organization is crucial for implementing its asset maintenance strategy. The assets' optimum performance can be ensured by adequately planning spare parts of the specific asset or equipment. Coordination among contract and procurement, asset maintenance, and finance departments is essential in streamlining MRO supply chain processing.

The MRO supply chain regarding spare parts management can be managed internally by the organization and outsourced to the vendors. Most operations-intensive organizations manage the MRO supply chain internally. Effective management of spare parts by asset maintenance managers enables the organization to achieve the best inventory control and material cost optimization. Outsourcing to an external agency or OEM vendor has its advantages like better inventory management and saving asset maintenance manager's time by achieving maintenance performance improvements. However, few organizations use the hybrid approach for accomplishing trade-offs between inventory management and material costs.

Often, asset maintenance managers overestimate the spare parts requirements to ensure nonavailability of the spare parts during breakdown maintenance. In the case of critical equipment, the spares' nonavailability leads to the stoppage of plant operations. The plant stoppage is a more embracing situation for the asset maintenance managers. Some of the operations-intensive organizations are restricting the spare procurement to minimize the material cost. Therefore, the organizations are now tending toward the spare parts management by OEMs. They enter into comprehensive annual contracts with OEM vendors for managing spares parts, supplying as and when required, and providing expert service. The technology developments enable to make such contract arrangements more effective. As discussed in Chap. 1, the IoT-enabled spare inventory control and just-in-time spare parts model are becoming the approach for today's organization. Such technology models provide the following benefits: optimum use of consumables, keep minimum shop floor inventory, less storage space, reduction in spare parts management interventions among the departments, right spare at the right time, and cost control.

Asset maintenance managers usually plan the spare parts based on OEM recommendations and their own experience based on equipment history. Despite the best planning by asset maintenance managers, spare parts management's challenges are spare inventory tracking, spares requirement in case of repeated failures, spare inventory according to spare part consumption pattern, obsolete equipment, and overstock of spare parts in anticipation of aging-related failures. For example, spares are procured for equipment like a compressor with due considerations of the factors mentioned above. There may be intimation from OEM that the particular compressor model has become obsolete. Then, the spares in stock may become redundant in case the compressor is replaced with another model. If the compressor is not replaced with a new model, there may be a shortage of spares. To tackle this issue, the asset maintenance managers adopt the local development of equivalent spare parts with the vendors available near the plant's location. Therefore, the recent developments in Real-Time Location Systems and Warehouse Management Systems could help to source the right spare at the right time quickly. Some of the operations-intensive organizations have adopted such recent developments in OR and MS applications.

The critical elements of spare parts management (Fig. 3.5) adopted by asset maintenance managers in operations-intensive organizations to optimize inventory management using OR and MS recent developments are spare parts supply chain cloud, Real-Time Inventory System, Warehouse Management System, RFID tracking system, and integration with ERP system.

Spare parts supply chain cloud is a service that enables a proactive approach for improving supply chain processes in the industry. The supply chain cloud service requires cloud infrastructure, cloud-based software applications, easy data access, integration with new technologies, and cybersecurity features to effectively implement MRO supply chain management. The key features of the cloud service are as follows:

- Better integration of spare parts supply chain management processes and delivers end-to-end visibility for quicker solutions in the organization.
- Spare parts data analytics and development of reports for inventory monitoring and control is an important feature.
- The cloud service must be agile enough to respond to stakeholder's changes and real-time requirements.
- It must use new technologies such as Artificial Intelligence, machine learning to automate the spare parts supply chain processes.
- Cloud architecture platform must be an open platform to minimize the upgrading-related changes and enable faster integration on new platform changes.

Real-Time Inventory System is an automated system that tracks from the spare part's entry into plant premises to the end, i.e., spare parts consumption. The complete automated process includes spares ordering, receipt of the spare part at

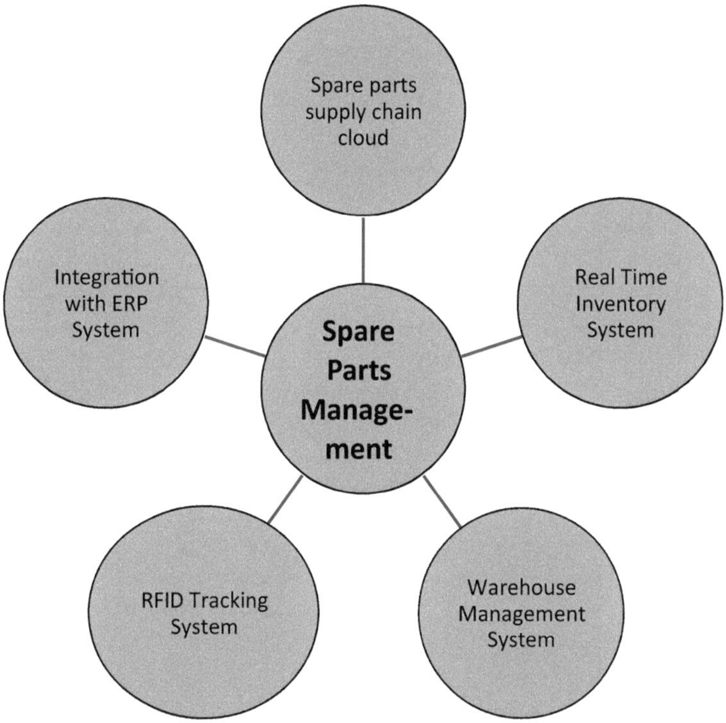

Fig. 3.5 Spare parts management: OR and MS application

stores, inspection, storage at a procurement department store or shop floor, retrieval on the requirement, issue of the spare part to the consuming department, and consumption of the spare. After consumption, automatically reorder based on the inventory level fixed for the particular spare item. Such automation facilitates the elimination of bottlenecks and avoids interruption and delay in spare parts management processes. In today's VUCA environment of business, spare parts workflow automation provides a competitive advantage to operations-intensive organizations.

Warehouse Management System is useful when the spares management is outsourced to OEM vendor or other spare parts management companies. The applications of such systems are given below:

- Identification and source of spare parts of equipment for overhauling are planned during the annual shutdown. After completion of the shutdown, return the unused spares back to the warehouse.
- Preparation of spare parts list as per preventive maintenance schedules makes such spare parts ready for ensuring timely availability.
- Work order management in case of emergency and planned jobs. For scheduled jobs, ensure the right spare at the right time to avoid any delay in maintenance jobs. Immediate availability of spare for emergency breakdown jobs.

- Centralized spares management across the organization is critical to optimize the spare parts inventory of assets/equipment.

RFID tracking system enables the asset maintenance processes in the following ways: spare parts monitoring and their visibility, effective spare parts management in a warehouse, plant stores and shop floor, minimizing lead time in the availability of spare parts for maintenance jobs, cost saving through optimum stock inventory, equipment shutdown reduction, and improving maintenance job quality. The RFID tracking systems provide real-time spare parts availability, auto-replenishment of spares tracking in inventory level, material movement traceability, and identification of instant geolocation of the spare part. The RFID tracking system helps asset maintenance managers maintain documentation of equipment-wise failures, eliminating the errors in spare parts management, enabling the PM schedule execution, and documenting spare parts specifications.

Integration with ERP system integrates the plant maintenance module of ERP system with the materials management module for better spare parts management in alignment with work order management to streamline the asset maintenance processes. The integration among the various processes such as work order receipt, spare parts issue from stores, and spare part consumption during the execution of work in the plant maintenance is crucial in improving the effectiveness of asset maintenance processes.

Key Takeaways

- A common thread in all the existing asset maintenance strategy frameworks/models is alignment between corporate strategies and asset maintenance strategies.
- Asset maintenance strategy's selection processes include the following significant elements: maintenance personnel experience, Original Equipment Manufacturer (OEM) recommendations, maintenance policy guidelines, asset criticality and availability, and process severity and complexity.
- Tacit knowledge available with maintenance practitioners is crucial in deciding on asset maintenance strategy at an organizational level in operations-intensive organizations.
- Decision Support System (DSS) models enable a choice among the asset maintenance strategies available for equipment or process for effective implementation of suitable maintenance methods to improve asset maintenance performance and optimize the operational and maintenance costs.
- DSS has an option to select between planned maintenance action and predictive model (which includes mathematical, neural network, and logical methods)-based maintenance action, which depends on the case-by-case basis of the asset condition or requirement.
- The selection of specific maintenance strategies in operations-intensive organizations is based on the following significant factors: Original Equipment Manufacturer (OEM) recommendations, industry standards, experience by the maintenance managers, and maintenance policies at an organizational level.

- The critical differentiator in maintenance strategy formulation is long-term/shutdown maintenance planning or short-term maintenance planning.
- One of the primary objectives of the asset maintenance strategy formulation is to identify suitable maintenance tactics or a combination of maintenance tactics for the asset/equipment/process.
- Critical equipment identification is the basis for specific maintenance strategy formulation for the assets undertaken to develop an asset maintenance strategic plan.
- Asset maintenance strategic plan is drawn based on internal norms such as maintenance policy and industry norms such as national/international standards.
- While formulating AM strategic plan due to care must be taken by maintenance managers to minimize the maintenance challenges they face in the past, suitable interventions should be incorporated in the plan based on their experience and history documentation.
- Optimization of total costs over the entire asset life cycle stages is one of the methods to achieve an organization's business objective while maintaining the availability of the asset to operations. Another method is to formulate a cost-effective AM strategic plan for each stage of the asset life cycle. The optimizations of maintenance cost across all the stages of the life cycle are crucial.
- A nine-step best maintenance practice includes the following steps: (1) build maintenance culture, (2) measure average maintenance cost/hour, (3) monitor production cost vs. maintenance cost, (4) timely availability of spares, (5) Computerized Maintenance Management System, (6) timely completion of PM schedules, (7) implementation of Reliability-Centered Maintenance, (8) implementation of Total Productive Maintenance, and (9) perform proactive maintenance.
- Managing the spare parts by asset maintenance managers enables the organization to achieve the best inventory control and material cost optimization.
- Outsourcing to an external agency or OEM vendor has its advantages like better inventory management and saving asset maintenance manager's time by achieving maintenance performance improvements. However, few organizations use the hybrid approach for accomplishing trade-offs between inventory management and material costs.
- The critical elements of spare parts management adopted by asset maintenance managers in operations-intensive organizations to optimize inventory management using OR and MS recent developments are spare parts supply chain cloud, Real-Time Inventory System, Warehouse Management System, RFID tracking system, and integration with ERP system.

References

Eti, M. C., Ogaji, S. O., & Probert, S. D. (2006). Development and implementation of preventive maintenance practices in Nigerian industries. *Journal on Applied Energy, 83*(3), 1163–1179.
Kelly, A. (2006). *Strategic maintenance planning*. Oxford University Press.

Li, L., Ambani, S., & Ni, J. (2009). Plant level-maintenance decision support system for throughput improvement. *International Journal of Production Research, 47*(4), 7047–7061.

McAllister, K., Armstrong, J., & Wilson, A. (1999). *Asset maintenance management, a guide to developing strategy and improving performance*. Conference Communication.

Muchiri, P., Pintelon, L., Gelders, L., & Martin, H. (2011). Development of maintenance function performance measurement framework and indicators. *International Journal of Production Economics, 13*(1), 295–302.

Pintelon, L. M., & Parodi-Herz, A. (2008). *Maintenance: An evolutionary perspective, complex system maintenance handbook*. Springer.

Salonen, A. (2011). Strategic maintenance improvement: Driving forces and obstacles. *5th World Congress on Engineering Asset Management*. Brisbane.

Umar, A. (2011). A framework for strategic planning in maintenance. *Journal of Quality in Maintenance Engineering, 17*(2), 150–162.

Velmurugan, R. S. (2016). Doctoral Thesis. In *Maintenance strategy and practices at a large gas utility company in India—A casestudy approach*. University of Petroleum and Energy Studies.

Velmurugan, R. S., & Dhingra, T. (2015). Maintenance strategy selection and its impact on maintenance function—A conceptual framework. *International Journal of Operations & Production Management, 35*(12), 1622–1661.

Chapter 4
Asset Maintenance Framework and Benchmarking

Abstract This chapter demonstrates the framework for maintenance strategy and practices for planning and execution at an organizational level, which is empirically tested through the research study conducted among operations-intensive organizations with the inputs from practicing asset/facility managers and engineers. The asset maintenance strategy processes are maintenance strategy formulation, maintenance strategy selection, and maintenance strategy implementation. These processes are interlinked and impact asset maintenance planning and execution in an organization. The enablers of asset maintenance strategy formulation are maintenance performance, planning and scheduling, materials management, human resources, Information Technology, employee empowerment, maintenance tactics, reliability analysis, maintenance policies/maintenance budget, and maintenance challenges.

The maintenance strategy selection process is based on Maintenance Decision Support System (MDSS) development suited to organizational requirements and execution of specific maintenance strategies for a particular process/equipment recommended by the equipment manufacturer or process licensor. Asset maintenance strategy processes must be aligned with the business strategies and operational strategies of an organization.

Implementation of maintenance strategy and maintenance practices is followed by the formulation of maintenance strategy and selecting a proper mix of maintenance strategy for the process/equipment/operations. It can be achieved using modern maintenance tools and techniques and software applications. An organization-wide integration of maintenance practices helps the organization implement the maintenance strategies efficiently across the organization. Further, elements related to maintenance practices toward benchmarking with a framework for benchmarking are also discussed in this chapter.

R. S. Velmurugan, T. Dhingra, *Asset Maintenance Management in Industry*,
International Series in Operations Research & Management Science 310,
https://doi.org/10.1007/978-3-030-74154-9_4

4.1 Asset Maintenance Strategy and Practices Framework

A research was conducted among the asset maintenance practitioners of operations-intensive organizations to understand asset maintenance strategies and current asset maintenance practices. The result of this detailed research is a comprehensive framework (Fig. 4.1) for asset maintenance strategy and maintenance practices planning and execution for operations-intensive organizations. The development of asset maintenance strategies, selecting the specific asset maintenance strategy for an asset/equipment, and implementing asset maintenance strategy with the

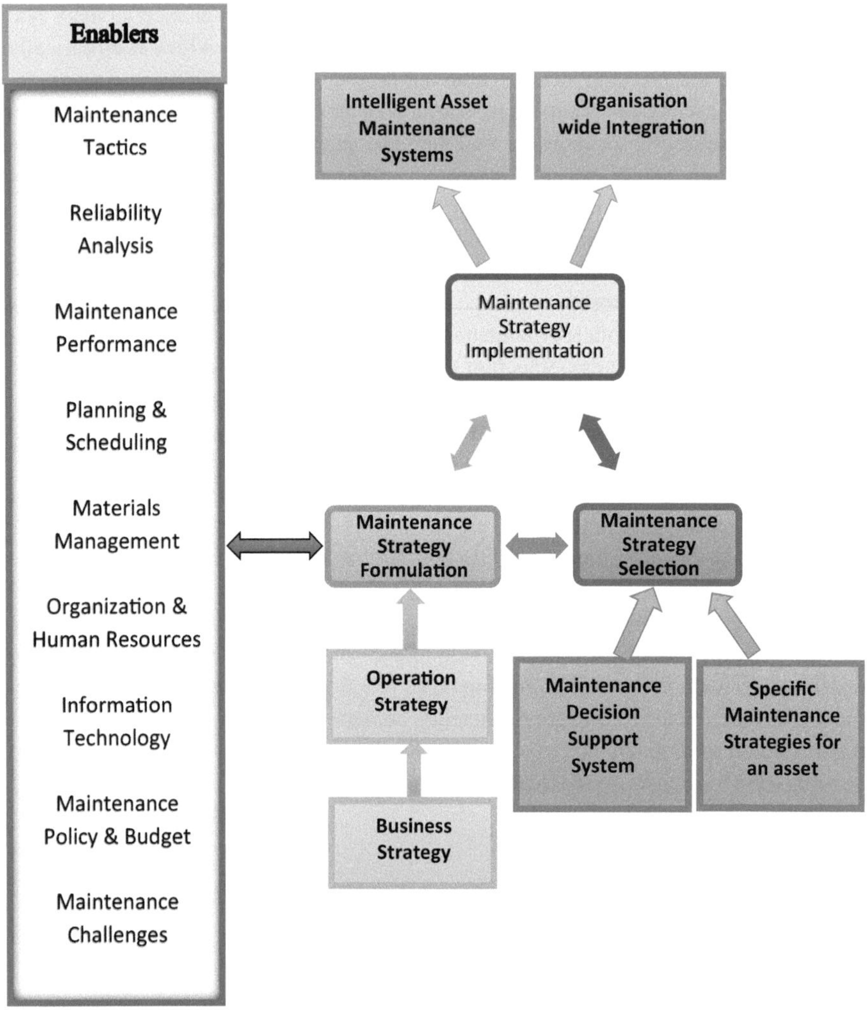

Fig. 4.1 Asset maintenance strategy and practices planning and execution framework for operations-intensive organizations

industry's best practices are interrelated asset maintenance processes. If all these processes are not appropriately managed, it is impossible to plan and execute an asset maintenance strategic plan at an organizational level.

In the suggested framework, the primary asset maintenance processes are asset maintenance strategy formulation, selection of asset maintenance strategy, and asset maintenance practices (Fig. 4.1), and all these processes impact the asset maintenance planning and execution. The maintenance strategy formulation is developed based on the process inputs from asset maintenance enablers such as maintenance policy and budget, organization and human resources, materials management, planning and scheduling, maintenance performance, Information Technology, maintenance tactics, reliability analysis, employee empowerment, and maintenance challenges. The detailed roles of these enablers regarding asset maintenance strategy and maintenance practices are explained in Chaps. 2 and 3. The maintenance strategy formulation is also required to be integrated with operations and business strategies. The formulated asset maintenance strategies provide guidelines for asset maintenance enablers of their effective management at an organizational level.

The outcomes of the maintenance strategy formulation are the inputs to the asset maintenance strategy selection. The asset maintenance strategies developed are evaluated using Maintenance Decision Support System (MDSS) designed based on organizational requirements and the specific asset maintenance strategies recommended by the Original Equipment Manufacturer of an equipment/asset or the process/manufacturing operation's licensor. The asset maintenance selection process includes developing an appropriate mix of asset maintenance strategies for any plant operations/equipment/asset. A modern asset maintenance approach such as e-maintenance and intelligent automation systems must be used to improve the asset maintenance implementation toward benchmarking of the asset maintenance.

The implementation of asset maintenance strategy and practices of a plant/ operational unit within the organization must be integrated across the organization in all the similar plants/operations units to leverage the resource optimization and standardize the asset maintenance processes. It is found from the research study that the asset maintenance managers in operations-intensive organizations are more concerned about asset maintenance challenges. The reason is that despite the best asset maintenance strategic plan, the assets/equipment failure happens, and the plan cannot be executed due to these challenges. Therefore, proactively including the preventive steps or emergency action plan during the asset maintenance strategy formulation is the key growth driver to asset maintenance management. Further, asset maintenance managers in operations-intensive organizations are adopting the asset maintenance strategies recommended by the Original Equipment Manufacturer (OEM) of an equipment or process/production licensor.

The combination of empirical findings from the research study conducted in operations-intensive organizations and the theoretical frameworks available in the literature resulted in a comprehensive framework (Fig. 4.1).

The asset maintenance strategy and maintenance practices planning and execution framework for operations-intensive organizations are described below with the empirical findings arrived from this framework. These findings can be checkpoints to understand the existing maintenance processes and be used as asset maintenance practitioners' guidelines.

4.1.1 Asset Maintenance Strategy Enablers

Commonly, maintenance tactics such as preventive maintenance, predictive maintenance, and unplanned maintenance are used in operations-intensive organizations. Predictive maintenance practices are preferred over preventive maintenance in the industry to maximize the availability of the asset/equipment and minimize unplanned maintenance since the production/manufacturing facility is operating continuously on a 24/7 basis. Few organizations follow maintenance tactics such as Total Productive Maintenance and Reliability-Centered Maintenance. It is evident from the research study that the asset maintenance managers in the industry agree to the implementation of the maintenance tactics such as TPM and RCM, which are more beneficial to improve the reliability in asset maintenance and achieve the best world-class maintenance practices.

Maintenance managers/engineers indicated that **shutdown maintenance** is a crucial maintenance tactic for the following reasons such as plant/facility is operating continuously in three shifts in a day to achieve the production targets, no standby equipment is available for the critical equipment used in plant operations, and maintenance activities are planned to be performed as per OEM recommendations. Therefore, most operations-intensive organizations use a shutdown maintenance approach, i.e., annual shutdown or longer duration shutdown in every 2–3 years of the process/manufacturing plant is taken for performing the planned maintenance activities on the critical assets/equipment.

Proactive maintenance is another maintenance tactic that uses asset maintenance practitioners' experience to decide on maintenance interventions in the plant/facility. The elements that affect proactive maintenance are maintenance personnel experience, limitations in operations, geographical area limitations, maintenance crew's culture, and top management support implementing chosen maintenance strategy to practice benchmarked maintenance parameters (Velmurugan, 2016). Such kind of maintenance approach is required for operations-intensive facilities/processes. The elements contributing to the proactive maintenance are shown in Fig. 4.2.

Proactive maintenance is a preventive strategy that aims for root cause analysis (RCA) of asset maintenance failures and avoids an emergency shutdown of operations caused by underlying equipment conditions. The primary objective of proactive maintenance is to predict equipment/asset failures in advance from the process conditions and eliminate the failure before it develops. A proactive maintenance plan helps operations-intensive organizations to find their unseen inefficiencies in asset maintenance. This kind of maintenance approach makes it possible to do maintenance only when need arises, and this decision can be inferred from the maintenance data.

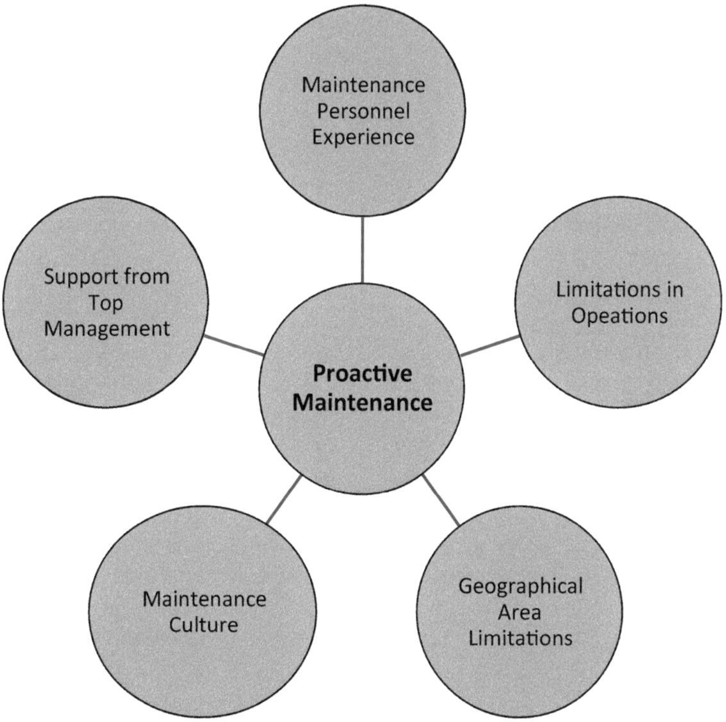

Fig. 4.2 Proactive maintenance

A proactive maintenance plan facilitates the maintenance practitioners to perform RCA and identify the root cause of the equipment failures, eliminate equipment failure before it happens, extend the asset life cycle, and ensure the asset's availability for operations. Integration of a proactive maintenance plan with the Computerized Maintenance Management System, focusing on failure modes, support from top management, suitable technological implementations, and upskilling and reskilling of maintenance workforce, is a recommended AM strategy for the operations-intensive organizations.

The proactive maintenance approach requires more investment and time than other standard maintenance approaches—the outcomes after implementing proactive maintenance yield cost saving and efficiency improvement in asset maintenance performance. Preventive/predictive maintenance tactics can influence proactive maintenance. On the other hand, proactive maintenance improves these maintenance tactics' efficiency and streamlines maintenance costs.

The following findings are related to the enabler maintenance tactics based on the above discussions:

Finding 1: *The use of predictive maintenance integrated with intelligent automation systems to a larger extent is a crucial enabling asset maintenance process for operations-intensive organizations.*

Finding 2: *A meticulous planning of shutdown maintenance with proper resources such as material, workforce, and OEM services is essential for asset maintenance management in organizations operating on a 24/7 basis.*

Finding 3: *Asset maintenance practitioners must adopt a concept of proactive maintenance to improvise the asset maintenance processes.*

In continuous operations and processing plants, equipment/asset is critical in operations since no standby equipment is available except few processes. In batch processing, long-term planning and scheduling are not required. Such a long-term asset maintenance planning approach is called shutdown maintenance or annual turnaround in the industry.

It is a planned shutdown of the production/operations of the plant/facility in the organization. Planned maintenance activities are carried out during the shutdown period, requiring resources such as workforce, material, and other related requirements. Sometimes, turnaround shutdown is extended more than the planned duration due to unexpected jobs on the asset/equipment under maintenance. A production loss amounts to a significant portion of the organization's concerned plant/operations' annual budget. A key success factor of the plant turnaround/shutdown is good planning. Asset/equipment that require maintenance or replacement are to be identified during the planning stage, and required resources are also to be arranged as per standard operating procedures. A baseline analysis of an operations plant provides a complete overview of the turnaround shutdown's scope and duration.

Finding 4: *A long-term planning and scheduling of asset maintenance activities are critical for operations-intensive organizations. It is more critical in continuously operating process plants than the batch processing plant.*

Box 4.1:
Total maintenance budget = $10 million.
Spares + inventory cost = $3 million
(i.e., 30% of maintenance budget).
If the spares and inventory cost are reduced by 10%, then
cost saving = $300,000, i.e., 3% saving from total maintenance budget.

In organizations, materials management is carried out through Enterprise Resource Planning software applications like SAP modules. SAP systems have a separate SAP MM module for materials management. Spares and inventory costs of asset maintenance account for 25%–45% of the total maintenance budget. Optimizing these costs is a challenge for maintenance practitioners since they are mostly involved in their major maintenance works like work order management and preventive maintenance schedules. Therefore, reducing spares and inventory costs leads to cost savings to operations-intensive organizations. An illustration (Box 4.1) shows that a 10% reduction in material costs resulted in a 3% saving from total maintenance cost. Therefore, integrating the Computerized Maintenance

Management System with the materials management system is important for optimizing the material cost in maintenance costs. The organizations that are using ERP-SAP applications could integrate their SAP materials management (MM) module with the SAP plant maintenance (PM) module seamlessly.

To achieve the cost saving in material cost, few steps to be taken by maintenance practitioners are as follows:

- A yearly report on expenditure trend (month-wise) on spares and inventory for asset maintenance is obtained from the materials management system or CMMS.
- A detailed analysis is to be carried out by the maintenance manager in terms of material cost against overall maintenance cost, equipment-wise number of preventive maintenance schedules executed, spare consumption pattern, unused spares for more than 3 years, sourcing cost due to emergency breakdown, the trend of consumables usage, and repeated consumption of a particular spare item.
- In case an organization has more than one plant/facility, then the above steps are to be carried out for every plant across the country or countries.
- In case of multiple plant locations, cross-analysis of reports is to be carried out in terms of duplicate spares, repeated usage of a particular material, extended inventory of critical spares, and procurement of spares from vendors across the plant locations based on the factors like similar spares that are procured from different vendors, cost comparison, and purchase lead time.

Based on the above-detailed analysis, the cost improvement measures are reduction in long inventory items for holding cost optimization, maintaining an optimum level of spares across the plant locations/equipment, reduction of unidentified spare inventory in shop floors, optimization of consumables usages, and maintaining warranty records for material consumption of OEM vendors. All such measures ultimately reduce the material cost at a plant/facility level and save maintenance costs.

Finding 5: *Materials management systems must be integrated with Computerized Maintenance Management Systems or asset management systems. All-in-one system like Enterprise Resource Planning (ERP) provides betterment in asset maintenance processes in operations-intensive organizations in the industry.*

Effective implementation of asset maintenance strategy requires thorough planning during the AM strategy formulation stage itself. Human resources like maintenance technicians, maintenance engineers, maintenance managers, and support contract workers are the backbone in implementing the best asset maintenance practices. The research outcomes related to crucial implementation points in terms of HR development in the asset maintenance team are the following:

- Identifying all human resource requirements in the asset maintenance function based on the defined job roles and standard workforce plan helps place the right man on the right job.

- Operations-intensive organizations must maintain the human resources' demographic information with their detailed professional profiling based on their qualifications, experience, age, country, etc.
- Organizations must plan human resources related to asset maintenance function in key performance areas such as knowledge, skills, competencies, aptitude, and learning and development.
- Identifying high-potential employees based on their asset maintenance roles and right job allocation in asset maintenance function and placing them in the right job to leverage their potential in achieving maintenance excellence are facilitating aspects of achieving maintenance excellence.
- Evaluation of maintenance personnel against identified maintenance competencies provides them individual development training in development areas based on their roles and job description.
- Maintenance performance measures are required to be linked with the maintenance personnel performance management system (PMS) and key result areas of the individual employee.
- The effectiveness of learning and development interventions in asset maintenance is to be understood, and accordingly, the interventions can be modified suitably.
- Capturing experienced maintenance managers' tacit knowledge is a critical factor in knowledge management of asset maintenance area and further retaining, disseminating, and capability building to retain such knowledge when an employee leaves the organization proactively.

The implementation capabilities depend on the organization's size, such as small, medium, and large, and further on the organization's types such as private, government, and public sector. Financial and top management support are also essential factors while implementing HR interventions at an organizational level.

Finding 6: *Integration of HR processes with asset maintenance processes is essential in accomplishing maintenance activities in a quality and effective manner. There may be some barriers due to organizational capabilities that require the attention of the top management. Lacking HR strength impacts asset maintenance processes.*

The operations-intensive organizations leverage Information Technology to the larger amount in executing asset maintenance activities at an organizational level. The diagnostics expert systems used in maintenance strategy implementation are generally based on the process complexity and type of equipment. Conventional diagnostic systems are limited to detecting failures such as system hardware, field instruments, and process conditions. However, the current smart diagnostic systems can predict failures by monitoring field equipment/instrument parameters trends and system data. Prediction diagnostic systems are possible due to intelligent field instruments and technological development in instrument hardware systems. Earlier, the maintenance team acts on intimation from process operators regarding the abrupt change in process conditions to troubleshoot equipment failures. Nowadays,

Table 4.1 Asset maintenance interventions

Asset type	Asset maintenance interventions
Pumps	Inspection, tuning, lubrication, repair, refurbish, replacement, testing
Motors	Oil change, belt inspection, power supply check, lubrication, repair, restore, replacement, meggering, testing
Compressors	Inspection, alignment checking, lubrication, repair, restore, replacement, testing
Transformers	Wiring checks, oil check, temperature check, power supply check, restore, testing
Valves	Inspection, leakage test, adjustment, repair, replacement, testing
Pressure transmitter	Wiring check, power supply check, leakage check-in tubes, calibration, repair, replacement

prediction-based systems accelerate instrument response time and system hardware failures well before reflecting failure in process conditions.

In these systems, the diagnostic function application is built-in to field instruments and equipment such as compressors, gas turbines, steam turbines, pumps, extruder, and bagging machines and their types. It also depends on the process complexity. The equipment/field instrument failure prediction can be shared with the process operator. Such information to the operator helps make quick decisions in operations based on predictive diagnostic alarms and ensures the plant's safety by preventing equipment/field devices' sudden failure. Therefore, such intelligent diagnostic systems help predict equipment trouble and carry out efficient maintenance practices that could result in maintenance excellence, operational excellence, and enhanced asset life cycle.

Finding 7: *Operations-intensive organizations leverage smart expert systems for performing asset maintenance diagnostics. The process complexity and criticality of the equipment are the primary elements for diagnostic systems designing.*

Maintenance policy provides guidelines for asset maintenance strategy formulation and selection in operations-intensive organizations. As stated above, maintenance policy guidelines are issued based on asset types, as shown in Table 4.1. This table provides examples of asset types and asset maintenance interventions performed in such assets/equipment.

The maintenance policy in operations-intensive organizations aims for maintenance excellence, i.e., optimization of asset life cycle cost, replacement of assets in case of safety and environment regulations, and obsolescence of equipment/asset. If the policy does not aim for such objectives, it may cause huge costs and low asset availability in the plant/facility. Therefore, maintenance policy must include some of the best asset maintenance practices such as proactive maintenance, prediction-based analysis of asset failures, and intelligent predictive maintenance systems.

The maintenance policy guides the organization to perform world-class asset maintenance to maximize business profits. Accordingly, it defines the best maintenance approach for the assets at an organizational level. Few elements to be considered while framing maintenance policy are asset performance, reliability, the life cycle of the asset, and the condition. It is found in the organizations that the maintenance policy is based on asset operating frequency, types of failure modes, and condition of the asset/equipment.

Asset operating frequency maintenance policy guidelines are prepared based on OEM recommendations. For example, compressor overhaul is to be done after 8000 h of running. However, the overhaul of a compressor is based on the compressor operating parameters in Condition-Based Maintenance policy. The maintenance of the compressor can be carried out even after 10,000 h. Another type of maintenance policy is based on failure modes in correlation with the asset's life cycle.

Finding 8: *Maintenance policy is developed considering the various aspects like types of equipment, asset aging, nature of the business, and industry standards in operations-intensive organizations. The maintenance policy's overall objective is to achieve maintenance excellence with an ultimate aim of operational excellence.*

The maintenance budget is of two types as revenue budget and capital budget. The revenue budget is prepared majorly for repair and maintenance of the asset/equipment of the plant/facility. A capital budget is prepared for replacing assets/equipment, revamping, retrofitting equipment, and new infrastructure. These budgets require equipment history data, failure analysis of equipment/asset, health assessment of equipment, and spare inventory. Long-term financial budgeting must consider the asset's complete life cycle under planning, prioritization, assessment of capital projects, and their operating expenditures. Further, due consideration must be given to underutilizing assets' cost impacts due to reduced service life, breakdown maintenance, and service troubles. The maintenance budget ensures the following deliverables: the required level of service, risk mitigation, and availability of the asset/equipment for continuous operations/production of the plant/facility.

While preparing the maintenance budget, the following key factors account for accurate budget preparation for the financial year:

- Comparison between actual expenditure and budgeted expenditure of last year's budget.
- Repair and maintenance requirements of all the assets/equipment in the plant/facility for preparing revenue budget for the financial year.
- Identification of capital requirements includes asset modifications, replacement of assets, and new procurement of assets/equipment to enhance the production capacity in alignment with the maintenance policy and maintenance strategy for preparing a capital budget for the financial year.
- Life cycle cost analysis of the asset/equipment to prioritize and assess the revenue and capital budgets.

Finding 9: *A meticulously planned asset maintenance budget is a critical success factor for effectively executing the chosen asset maintenance strategies in operations-intensive organizations.*

Maintenance challenges is the hidden asset maintenance contributing factor highlighted by the asset maintenance managers in operations-intensive organizations during the research study. These issues are categorized into issues from various areas such as technical, contract, internal, and stakeholders of the plant operations, but the issues identified are not limited to these only. The maintenance challenges vary from organization to organization and plant to plant since the issues about maintenance challenges depend on various factors in the plant environment. For example, in gas/oil pipelines, **unexpected issues** are more since stakeholders' involvement is more like farmers, contractors, local government policies, and political leaders. In petrochemicals/refineries, **standardization of maintenance processes** can be a challenge due to the complexity of asset maintenance processes in such operational plants.

In operations-intensive organizations, achieving operational excellence and maintenance excellence is also one of the maintenance challenges. A well-structured asset maintenance strategic plan is the enabler to achieve operational and maintenance excellence in the organization. The asset maintenance team leader is responsible for taking care of maintenance excellence aspects in managing assets. Therefore, there is an immense requirement for long-term and short-term tradeoffs of operations targets, asset integrity, safety and risk obligations, and maintenance cost.

One of the challenges to the maintenance managers is to improve the plant/facility's operational excellence by ensuring the availability of the asset/equipment. Asset availability could be achieved by the suitable maintenance strategy implementation for an asset/equipment with the appropriate balance of maintenance tactics such as preventive, predictive, and proactive maintenance. This balanced approach maximizes asset utilization over the medium to long term over the asset life cycle while ensuring the optimum maintenance cost's asset integrity. To achieve this object, maintenance managers must have in-depth knowledge about assets under maintenance that includes various factors such as operational environment, asset capacity utilization, and OEM recommendations. This knowledge needs to be transformed into effective asset maintenance strategies and detailed maintenance job plans that enable equipment-specific maintenance interventions such as inspection, lubrication, replacement, and testing.

Despite implementing the best asset maintenance strategy, unexpected asset/equipment failure is a ubiquitous reality. The asset maintenance managers must be ready to manage such kind of contingency management whenever it happens. The maintenance team's quick reactive maintenance capabilities should be an integral part of any well-balanced maintenance strategy plan. Maintenance managers need to respond and manage the situation with the necessary support availability of spares, equipment service support from OEM, troubleshooting plan, and resolving the trouble at the earliest possible. Further, balancing the maintenance cost on technology, skilled workforce, critical spares, and OEM services is the significant responsibility of maintenance managers while considering the asset/equipment's productive value. Maintenance costs include all types of costs such as fixed, variable, direct, and indirect costs aside from depreciation and other tax implications.

A challenge for maintenance managers is to understand the business objectives at an organizational or unit level where the asset is used for production/operations and maintain the impact of maintenance-related choices and trade-offs in long-term and short-term asset maintenance planning.

Another area of challenge for maintenance managers is spare inventory optimization. It is always a choice of the maintenance team that keeps all the spare parts and materials of an asset so that maintenance activities can be carried out without any delay in the breakdown of asset/equipment. Simultaneously, carrying such spares stocks in the long term for only expecting the asset failure leads to a vast inventory holding cost. Otherwise, the nonavailability of spare may lead to a long shutdown of the plant/facility that results in operating profit loss. A trade-off between these two situations, unnecessary inventory and no inventory, is the challenge for maintenance managers. Therefore, meticulous maintenance planning and scheduling help to resolve such a scenario.

Even though asset maintenance is a physical activity in maintaining assets for ensuring continuous production in operations-intensive organizations, maintenance execution is a human resource management process. One of the challenges of maintenance managers is ensuring the correct balance among the maintenance processes and workforce utilization, which requires people management, communication among the team, negotiation among all the stakeholders, and accurate planning of maintenance activities. Such collaborations among the teams are a competency an asset maintenance manager must possess or develop. Therefore, maintenance managers' major role is to facilitate the communication channels and integrate work processes with all departments such as engineering, operations, procurement, stores, safety, and human resources. Further, it must align with the maintenance organization itself, OEM support providers, suppliers, and vendors. The maintenance team leader is also part of the organization's top management team to make the business, operational, and financial decisions at an organizational level. Modern maintenance management systems facilitate developing a collaborative working digital platform where data related to maintenance activities can be shared on a real-time basis.

Standardization of maintenance practices is a challenge for maintenance managers since most operations-intensive organizations have multiple plants/facilities located within the country or across the countries. Collaborative working could be a challenge due to the geographical separation of process plants/facilities, limited corporate guidance, and absence of standards and coordination among the maintenance groups. Various asset maintenance strategies and maintenance approaches to similar and identical assets/equipment contribute to variability, operational risks, and higher costs. Leveraging the knowledge available and building uniform asset management teams within the organizations are the misalignment of organizational objectives across the plants/facilities. To mitigate such a disadvantageous scenario, adopting suitable asset maintenance standards is critical to maximizing Return on Asset (ROA). Therefore, standardization of the maintenance work processes, skills, materials management, and knowledge dissemination of work practices among the plants/facilities is the main objective of the asset maintenance strategy and practices.

Standardization does not mean that all production plants/facilities must do asset maintenance strategies and practices in the same way. Specific asset maintenance strategies and practices are always essential, but standardization helps maintain consistency in asset maintenance practices and data across the organization's plants/facilities.

Organizations are usually discouraged from developing asset maintenance standardization due to the following reasons: (1) the investment of time, cost, and other resources and (2) the experience and expertise required to implement the standardized practices. However, such standardization of asset maintenance practices is beneficial in the following directions such as consistency and collaboration among the plants/work centers, different sections within the plant, and operations and maintenance teams; reduction in variability among the various maintenance processes; improvement in safety, health, and environment systems; consistent maintenance performance measurement; identification of baseline performance of the maintenance function; and establishment of the continual improvement process. The baseline identification makes it possible for maintenance practitioners to swiftly visualize improvements about the asset maintenance processes that can be evaluated and tested. A baseline survey also helps maintenance managers combine the process and evaluate all the corresponding variables before deciding whether any changes must be incorporated into the processes. The detailed baseline survey process for developing an asset maintenance strategic plan is described in Chap. 6.

Standardization is not limited to maintenance processes, as it encompasses business processes such as standardization of all datasets, job roles and responsibilities, organization structure, and performance measures and targets at an organizational level. Almost every aspect of asset maintenance management must be standardized. Documentation is an essential aspect of the standardization process. Without the documented asset maintenance processes and uniformity in work procedures across the plants/facilities, it is impossible to achieve standardization in asset maintenance. Organizations need to focus on the following significant areas: top management and organization structures, critical datasets, asset maintenance practices, Information Technology, and asset performance management. Internal and industry norms guide the standardization at an organizational level.

Finding 10: *Asset maintenance challenges are specific to the organizations based on the type and complexity of plant operations, processes, and equipment. Identification of organization-specific challenges and their mitigation action plan is the deciding factor for asset maintenance excellence.*

Finding 11: *Standardization of all asset maintenance processes is the major challenge for asset maintenance practitioners in operations-intensive organizations. Understanding peer practices and adopting industry best practices are the elements that facilitate the standardization process.*

Overall, the asset maintenance enablers play a crucial part in asset maintenance processes and the development of asset maintenance strategies for asset-intensive operations. All these enablers are described in Chap. 2, and their correlation with asset maintenance strategy formulation is discussed in Chap. 3. Therefore, asset maintenance managers must consider asset maintenance contributing factors while developing the asset maintenance strategic plan. These factors are the basis for the choice of maintenance strategy and its execution with the best asset maintenance practices.

Finding 12: *Ensuring standardization of all asset maintenance processes related to asset maintenance enablers (Fig. 4.1) across the organization is the lifeline for asset maintenance practitioners in operations-intensive organizations. Without such support, it is impossible to achieve asset maintenance excellence.*

4.1.2 Asset Maintenance Strategy Processes

Asset maintenance strategy processes include maintenance strategy formulation, maintenance strategy selection, and maintenance strategy implementation (Fig. 4.1).

Maintenance Strategy Formulation

The asset maintenance enablers discussed provide support for asset maintenance strategy formulation. The strategy formulation must integrate with the objectives of operations and corporate at an organizational level.

In the current VUCA business environment, the strategic alignment of corporate objectives within the organization in all the business processes is the deciding factor for profitable business operations in operations-intensive organizations. Asset maintenance is also treated as one of the business processes in the organization. The alignment of asset maintenance processes and their objectives with the corporate objectives for achieving the organization's overall profitability is one of the organization's critical success factors. In the last decade of the industry, this is a significant thrust point and emphasized in all asset maintenance strategy frameworks (Chap. 3). However, it is a major challenge to operations-intensive organizations.

In the research study, the asset maintenance managers/engineers indicated the importance of percolating targets from the corporate level to an individual level. It must be a structured and participative process to align first to the corporate-level objectives with the plant-level objectives. Once this is done, the plant-level objectives are to be aligned and developed with the department-level objectives, i.e., operational objectives and asset maintenance objectives. The next logical step is, for example, aligning the asset maintenance objectives with the individual asset maintenance manager or engineer. Even though the top leadership takes steps to achieve such alignment to achieve the smooth execution of corporate-level strategies into functional strategies, much improvement is required. Therefore, the alignment part is to be taken care of while developing the asset maintenance strategic plan at an organizational level.

Finding 13: *Alignment of asset maintenance objectives with the corporate objectives is one of the critical success factors of operations-intensive organization. Further, asset maintenance targets are decided in line with operations/production targets to achieve the plant/facility's operational excellence.*

Organizations' business strategies are developed with a long-term perspective, while planning and executing any functional activities are viewed as part of a short-term perspective. Such traditional thinking, i.e., parting of business strategy from functional strategy, means that the potential benefits derived by aligning and managing these strategies are either ignored or not fully exploited.

The challenges in integrating business strategy with the asset maintenance strategy in the operations-intensive organization are not completely understood: the approach, path, and level of alignment it takes to achieve the proper integration and optimum asset maintenance performance. The organizations do make a wrong understanding that the alignment of strategies happens on their own, in addition to the mistakes mentioned above of distinguishing between strategic business objectives as long term and functional asset maintenance objectives as short term. At the plant/facility level, the organizations misconceived that only maintenance performance improvement is the primary objective of asset maintenance management.

The maintenance managers presume that the maintenance strategies align naturally with the corporate objectives. Such misconceptions lead to failure in implementing corporate business objectives at a plant/asset level and inefficiency in maximizing the organization's performance. A practical and systematic approach is essential for asset maintenance strategy implementation at an organizational level. This approach must help the organization's interventions to ensure the long-term strategy at the business level and short-term strategy at the functional level. These interventions must be supportive of each other and can be used for planning and execution.

Asset maintenance managers can adopt several efforts to connect the gaps between corporate objectives and asset maintenance objectives. The steps toward this direction are given below:

- Development of an overall organizational level strategy for asset maintenance management should include risk management, regulatory requirements, operations and maintenance strategies, and upgrade requirements of assets.
- Define the roles and responsibilities and proper organization structure of asset maintenance functional area by clearly mentioning the activities related to production, planning, reliability, integrity, safety, and maintenance of all assets at the plant level.
- A systematic approach for identifying gaps in the existing asset maintenance practices and following the best maintenance practices is described in Chap. 3. These practices help achieve efficient asset maintenance management by aligning, integrating, and sharing resources across the country's plants/facilities or even across countries.
- Plan and schedule all functional strategic activities related to asset maintenance based on overall strategy at an organizational level. For each activity in the area of

asset maintenance management, targets, scheduling, levels, unit-level objectives, and resource requirements must be incorporated for standardization across the plants/facilities.

- Long-term strategic business objectives are to be aligned with the asset maintenance overall strategy for making decisions such as replacement of assets, procurement of additional equipment/systems, and service requirements from OEMs to support plants' continued operation.
- Deviation in achieving milestones can be identified by monitoring overall asset maintenance strategic plan implementation regularly. Action points to mitigate such deviations in the implementation plan in asset maintenance are to be prepared. Accordingly, the plant/facility's continual operation must be ensured by keeping assets/equipment availability.
- Maintenance performance measurement helps to identify variations between the expected and actual results in asset maintenance activities. Standardization of maintenance performance metrics across the organization is beneficial to achieve overall maintenance objectives.

Finding 14: *A systematic approach for bridging the gaps between asset maintenance strategies and corporate strategies must be in place to achieve overall organizational excellence.*

Effective implementation of asset maintenance strategies must adopt a comprehensive, tactical, and systematic approach focusing on integrated planning so that the large deviations in long-term strategic plans aligning with functional strategies and with the actual performance can be minimized. To reach such strategic objectives, organizations must increase visibility across all functional areas and integrate the related processes. Therefore, successful implementation of strategies does not happen just like that, and it requires a lot of effort by top management and the asset maintenance managers in the organization. Incorporating a structured approach allows organizations to connect all the asset maintenance processes and eliminate silos among the departments. Further, it enables the organizations to benefit from such synergies within the various functional areas and deliver on corporate business strategies that lead to optimum performance assets in their entire life cycle.

Maintenance Strategy Selection

It is evident from the research findings that operations-intensive organizations are not applying mathematical models to the full extent in selecting asset maintenance strategies. Such models in the Condition-Based Maintenance method help the maintenance managers implement the scheduled maintenance technically and in economically feasible way. Datasets in condition monitoring are used for both diagnosis and prediction purposes. Both diagnosis and prediction are crucial for a Condition-Based Maintenance program.

The automation of Condition-Based Maintenance depends on the number of smart sensors available in the maintenance processes and field data capturing. The predictive data using mathematical modeling help maintenance managers make optimized decisions on asset maintenance based on the time horizon of the asset/equipment condition available to perform asset maintenance. Such predictive tools also indicate the risk of the asset/equipment's possible failure from low- to high-risk grading scale. Accordingly, asset maintenance decisions can be made considering a particular asset/equipment condition and the organizations' risk appetite.

The tools and techniques used by asset maintenance managers are inadequate for deciding on the suitable asset maintenance strategy selection, formulation, and implementation. It is essential and needs the current business scenario to upgrade predictive asset maintenance tools for asset maintenance decisions. It is evident from research findings that the recent developments in asset maintenance management demonstrate that the data necessary for condition-based predictive analysis tools are already available in most organizations in the industry. It is now time to apply the latest mathematical models to take advantage of data more comprehensively to decide on asset maintenance strategy and adopt the best maintenance practices.

The long experience of the asset maintenance managers in the similar process plant or operations, the recommendations from OEM for developing asset/equipment-level maintenance approaches, internal norms of organizations like asset maintenance policy and standard operating procedures, asset/equipment criticality and availability, types of operations like complex or severe, and industry norms such as standards and regulatory policies are drivers for asset maintenance strategy selection at an organizational level in operations-intensive organizations. The detailed discussion is provided in Chap. 3.

Finding 15: *Operations-intensive organizations must leverage mathematical models and asset maintenance optimization models to the larger extent for asset maintenance strategy selection.*

Finding 16: *The drivers for asset maintenance strategy selection are asset maintenance manager's knowledge and experience, OEM recommendations, internal norms at an organizational level, and industry norms. These facilitate to development of a specific asset maintenance strategy for an asset/equipment.*

Maintenance Strategy Implementation

It is evident from the research that operations-intensive organizations are adopting and implementing modern asset maintenance methods such as e-maintenance, intelligent automation systems, and computer applications like CMMS and EAMS for efficient implementation of the asset maintenance strategies. If a new CMMS is introduced, organizations must complete all asset maintenance management practices and incorporate solutions for all the challenges faced by maintenance managers in the new system. The investment in the system can be evaluated through a measure of Return on Investment (ROI). The measurement is defined as follows:

$$\text{ROI on CMMS} = \frac{[\text{Cost savings in operations and maintenance} - \text{CMMS investment}]}{[\text{CMMS investment}]} \times 100.$$

The basic asset maintenance information is available readily with maintenance managers through CMMS. The decisions can be taken quickly based on timelines and more accurate asset information. Better planning can also be carried out from inferred information from CMMS. Therefore, CMMS delivers real-time cost visibility and provides other benefits such as improvement in energy consumption; achieving regulatory compliance in safety, health, and environment (SHE); and timely implementation of internal audits related to asset maintenance systems. Implementation of CMMS achieves cost benefits in the following areas:

- Speeds up the maintenance work order processes so that the maintenance job on a particular asset is performed as quickly as possible and puts back the asset into operations, i.e., cost saving by reducing the lead time on the maintenance of an asset.
- Modern maintenance management systems help fix failures before it happens and avoid such failures by performing pre-planned maintenance interventions. If the failures are not taken care of timely, it can lead to major failure and cause the shutdown of asset/equipment/operations, i.e., the pre-planned maintenance feature in CMMS saves costs by avoiding major failures.
- Vendor payments that include third-party contract services and OEM expert services are part and parcel of the asset maintenance cost. CMMS enables to decide on the best service provider, assess the value of availing vendor services, and plan for smarter deployments of such services based on historical data available in the system.
- CMMS monitors asset performance during the complete asset life cycle and provides timely information on wear and tear of asset/equipment components, patterns of failure modes, and predictive analysis of asset healthiness. Asset replacement decisions are taken based on a CMMS report on a particular asset/equipment. Pareto reporting is one example of CMMS reports, which facilitates informed decisions based on maintenance interventions against the asset's expected performance. An asset maintenance report derived from CMMS helps decide on the replacement of assets at the right time after utilizing the asset for its maximum life.
- CMMS could also help maintain optimum spare inventory based on historical data on stock levels, usage of the specific spare material, and spares' criticality. In consumables and identified spare items, stock reorder points can be set in CMMS and provide alarms on low stock.

In addition to the CMMS, the intelligent automation systems that emerged out in Industry 4.0 are the lifeline of implementing asset maintenance processes in operations-intensive organizations. One of such systems is Intelligent Predictive

Maintenance. A detailed discussion is provided on the IPM framework and its implementation strategy in Chap. 5.

Nowadays, another improved version of CMMS is available, i.e., Enterprise Asset Management System (EAMS). It provides a range of capabilities that focus on asset maintenance, monitoring asset maintenance plans, managing the asset maintenance activities, and forecasting the maintenance resource requirements to leverage the maximum utilization of assets in organizations. EAMS facilitates combining the asset maintenance management processes and successfully executes the asset maintenance strategic plan during the complete asset life cycle period of an asset/equipment in the plant.

Finding 17: *Implementation of modern maintenance techniques and recent technological developments in the asset maintenance domain is inevitable for operations-intensive organizations. The top management of the organization must understand their existing capabilities and provide a suitable environment for asset maintenance managers to implement new methods and techniques for improving their business profits.*

Enterprise Resource Planning (ERP) system applications provide business process management modules that manage its complete datasets across the business. This system covers a whole bunch of capabilities such as human resources, plant/asset maintenance, contracts and procurement, finance and accounts, sales and distribution, learning and development, and project management, and it can replace the existing individual software available and integrate a variety of other types of software applications in the organization. The ERP application integrates all the asset maintenance processes through the plant maintenance module and further with the materials management module for ensuring the proper spare parts management of an asset/equipment. The plant maintenance module's functionalities include planning and scheduling maintenance schedules, creating job requests, issuing job permits, asset/equipment history maintenance regarding details of the maintenance jobs performed in a particular equipment/asset, and maintaining standard maintenance instructions. The datasets available in the plant maintenance module are collected and analyzed at the organizational level. The required asset maintenance process modifications and maintenance policy review are carried out in operations-intensive organizations based on the analysis.

The Enterprise Asset Management (EAM) system explicitly focuses on asset maintenance management, asset maintenance monitoring and control, and asset maintenance excellence. However, the ERP system focuses on all the functions of the business. Therefore, selecting between EAM and ERP is based on the following aspects: whether organizations are asset focused, what kind of business software is being used in the organization, what is the size of the business, and what types of asset maintenance systems are available. Typically, organizations debate that it is better to include Enterprise Asset Maintenance as part of Enterprise Resource Planning or to have a separate EAM system. The benefits of integrating EAM/CMMS with ERP are given below.

EAM/CMMS system provides information regarding all the assets and ensures efficient utilization of assets/equipment over their life cycle, and finally, the system helps optimize asset availability and reduce costlier breakdowns. In the case of integration with the ERP system, organizations can determine how asset management impacts their financial objectives. An integrated system can handle everything from asset maintenance scheduling and planning to depreciation and capitalization of the assets/equipment in the ERP system comprehensively. Organizations having both systems are performing some of the tasks in both the systems. Therefore, duplicate tasks or information mismatching between the systems can be avoided by streamlining the asset maintenance processes by integrating the EAM system into an ERP system.

Box 4.2:
Total asset value = $10 million.

Estimated value addition by implementing ERP systems over 2–3 year = 10%.

Profit in business where the assets are being used = $1 million (i.e., $10 million * 10%).

Integrating EAM processes with the ERP processes enables the organizations' assets to become more profitable by monitoring and controlling systems and procedures in the ERP system. Such integration helps ensure asset availability during its life cycle and maximize Return on Asset (ROA). Industry experience from this research study says that ERP integration with the CMMS/EAM systems provides additional value in total asset value by maximizing asset availability. The 10% increase of total asset value can be achieved in case the assets are monitored and controlled using the ERP system over 2–3 years. An illustration to explain this argument is shown in Box 4.2.

Further, ERP system integration with EAM/CMMS can make asset maintenance processes more efficient in several ways. They are the automation of critical tasks, minimizing paper-based office administration, quality checks of assets/equipment, optimizing the spare inventory and consumables, proper implementation of asset scheduling and documentation, and integrating and standardizing asset maintenance activities across the plants/facilities of an organization. ERP and EAM system integration into a single ERP system at the organizational level fulfills the objective of "Administer assets and financials in one place." While the EAM system manages the physical assets and the ERP system manages the financials effectively and efficiently, merging these two systems with effective integration, organizations can leverage the asset maintenance management to impact the bottom line.

Finding 18: *Enterprise Resource Planning systems, as a single integrated system encompassing all the business processes, including asset maintenance processes, are used by many organizations. The dedicated systems of asset maintenance such as Computerized Maintenance Management Systems and Enterprise Asset*

Maintenance systems are also deployed in the organizations. The major challenge is to leverage the full potential of such systems. Many organizations are utilizing 50%–70% capacity of ERP systems. The solution is skill building of ERP systems of employees who use ERP applications in their day-to-day jobs.

An organization-wide integration enables the organizations to standardize the asset maintenance activities and testing procedures across all the plants/facilities. Such standardization drives organizational efficiency and achieves global standards compliance. The driving factors of organization-wide integration are given below:

- Various plants/facilities usually do not share their experiences, best practices, asset maintenance processes, and systems. That kind of silo approach can be avoided using enterprise-wide applications like ERP, facilitating information and data sharing among the plants/facilities across operations-intensive organizations.
- In case any modification or new addition in asset maintenance practices in the system reflects across all the plants/facilities, if the organization already follows organization-wide integration, it leads to quick scalability of the modified process and aids for business growth.
- The top management needs to communicate effectively to all the functional departments across the plants to achieve the corporate business objectives. The company-wide integration helps standardize performance measurement metrics such as key performance indicators (KPI), monitoring reports, budgeting, cost optimization measures, and asset utilization cost. Therefore, managers can connect and learn from one another to improve asset maintenance management practices' effectiveness.
- As stated above, standardized performance measurement metrics drive benchmarking among the plants/facilities that lead to competition among the plants in achieving asset maintenance excellence. Further, common problems can be resolved together in a team.

Finding 19: *An organization-wide integration of asset maintenance strategy-related processes enhances the best asset maintenance practices among the plants/facilities to maintain operational excellence.*

Case on ERP Implementation

Business objective: An Indian manufacturing company in the operations-intensive industries decided to replace various business management systems with a combined Enterprise Resource Planning (ERP) software. After thorough deliberation within the company management, it was decided to select the SAP system as company-wide integration software.

Solution: The subject company planned to implement the selected SAP system across the organization's plants/facilities across the country. A detailed implementation plan was prepared, and a separate corporate ERP implantation team was set up in 2017 to integrate various software systems available within organizations. The initial implementation was carried out with the help of SAP

(continued)

authorized vendors. Further, the system's maintenance for attending to day-to-day problems is carried out by its ERP team.

Outcomes: The new SAP ERP system helps implement comprehensive business processes across the various functions such as finance and accounts, human resources, plant maintenance, training management, sales and distribution, and project management. A separate module for all these functions has been developed and customized to the company's specific requirements. A successful deployment of asset maintenance strategies has also been implemented across the organizations in various production plants. This implementation of SAP systems provides benefits in asset maintenance, such as maintenance cost analysis and control at the company-wide level, automated maintenance scheduling and planning, extended asset/equipment life cycle, and support to specific maintenance strategies for a particular asset/equipment at the plant level in the organization.

4.2 Asset Maintenance Benchmarking

The benchmarking process is evolutionary, and it can be classified as internal benchmarking, industry benchmarking, and best-practice benchmarking (Wireman, 2010). *Internal benchmarking* compares asset maintenance practices among similar process plants or units/departments within the organization. This type of benchmarking can lead to incremental improvements in the maintenance processes and cannot make a breakthrough in the processes. The Return on Investment (ROI) from such improvement could be a value addition for the organization. *Industry benchmarking* is a comparison of asset maintenance practices among similar organizations or competitive organizations within the industry. In this type of benchmarking, the improvements are incremental and slow phase.

In contrast to the benchmark practices mentioned above, the *best-practice benchmarking* compares asset maintenance practices between the industry's indisputable leaders. The indisputable leader may be from similar industries across global locations. This type of approach could provide breakthrough improvements for the organization. Therefore, the plant/facility studies the asset maintenance processes outside its industry and modifies or adopts improved asset maintenance processes that can make a quantum lead in asset maintenance performance compared to its competitors. The early adopter or adaptor plant can save substantial maintenance costs and aggressively help achieve the organization's business objectives.

The operations-intensive organizations aim to achieve asset maintenance excellence by innovating solutions to improve the equipment available to critical operations. The criticality of operations, environmental conditions, asset maintenance practices adopted by an organization, value to the stakeholders, and many other factors impact benchmarking in asset maintenance management. Benchmarking in asset maintenance permits the asset maintenance managers/engineers to assess their existing asset maintenance practices and processes with the best practices adopted by the peer organization in the industry/across the industry. Based on this assessment, the objective is to develop a sustained developmental plan for execution at an organizational level to benefit from peer practices.

A standard method of benchmarking the asset maintenance process is comparing the parameter-wise values with key performance indicators' average values decided for asset maintenance processes (Komonen et al., 2011), i.e., the availability of a natural gas pipeline transmission system of an organization is 98%. This value needs to be compared with the average value of the organizations in the industry. It is a basic method; however, various qualitative and quantitative approaches are available to benchmarking asset maintenance. Qualitative methods provide an in-depth study of the industry's available organizations' information to arrive at detailed benchmarking data. Quantitative methods are more logical, collecting datasets and reaching final benchmarking by analyzing the collected data using statistical tools and techniques.

4.2.1 Benchmarking Framework

The benchmarking must always identify and envisage prospective value additions in the asset maintenance processes. It must help develop an improvement plan with the industry's maintenance practitioners' corrective actions for implementation. Authors (Valkokari et al., 2016) have developed value-based benchmarking for the mining industry. The outcome of the benchmarking is majorly dependent on the quality of the data collected from the industry. Further, data analysis is also an essential aspect in deciding the benchmarking in the organization. Therefore, the results must be validated by all the stakeholders to improve data reliability. Since the data collection is more critical in benchmarking exercise, the following points are to be accounted for:

- The organizations chosen for data collection must be similar within the industry.
- The other diverse organizations can be considered only for cross-validations of benchmark parameters.
- While collecting data, the developmental plan for the organization under study needs to be kept in mind.
- Among the plants/facilities, a comparable subgroup is to be prepared for processing units, pipeline systems, compressors, and gas turbines among the plants/facilities.
- The organizations' asset maintenance environments are to be collected in detail considering the environmental factors like the asset maintenance team's competency, maintenance infrastructure available for execution of asset maintenance activities, details of equipment's asset life cycle stage, and maintenance policies.

In line with the above, a questionnaire has to be developed for data collection and then data analysis to be carried out based on the collected data. The benchmarking exercise outcome is arriving at benchmarked data for every key performance

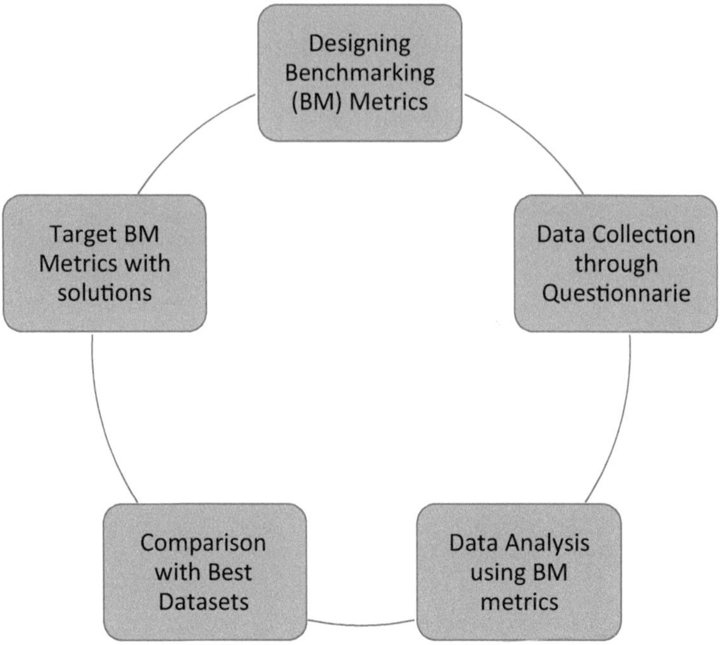

Fig. 4.3 Benchmarking framework for asset maintenance

indicator taken for the study. Further, the development areas have to be identified based on these benchmarked data for an organization under study and the appropriate improvement solutions. An asset maintenance benchmarking framework (Fig. 4.3) is discussed in detail below.

The benchmarking framework's first step is designing the benchmarking metrics (key performance indicators) regarding asset maintenance management for a particular plant or process or equipment as the case may be. The designing stage must involve all the stakeholders like asset maintenance and operations departments, top management, Original Equipment Manufacturers, and other key stakeholders. The brainstorming technique can be used to decide on the benchmarking metrics for asset maintenance activities at an organizational level. The second step is to develop a data collection questionnaire based on the metrics designed that covers all the relevant information to be collected from outside organizations. The data collection can be fruitful for accurate data analysis, i.e., the third step in the benchmarking framework. In this step, statistical tools are used to analyze the data, and the basis for data analysis is the benchmarking metrics developed in the first step.

The next logical step in the benchmarking framework is comparing the best asset maintenance datasets with the organization's datasets under study. The last step is setting target values for the benchmarking metrics and providing solutions for improving the development areas about asset maintenance management. The benchmark process does not end here if any new benchmark metrics emerged from the benchmark study; such metrics must be incorporated in the organization's list benchmark metrics under study. This benchmark helps to compare the best organizations' asset maintenance environment and improvised benchmarking metrics and targets to achieve maintenance excellence.

4.2.2 *Benchmarking Metrics*

Based on the research study conducted among operations-intensive organizations, a set of benchmarking metrics (Table 4.2) are developed for asset maintenance management. This template can be used as a reference by the asset maintenance managers/engineers of operations-intensive organizations in the industry.

Benchmarking metrics mentioned above must identify areas of development in asset maintenance management at an organizational level. In case any metric target is

Table 4.2 Benchmarking metrics for asset maintenance of operations-intensive organizations

Metrics	Method	Target range (%)
Plant/equipment availability	[Total running hours – (planned and unplanned down time)] × 100	94–99
Preventive maintenance schedule compliance	[Completion of scheduled jobs in hours/total hours scheduled] × 100	>90
Breakdown maintenance percentage	[Breakdown maintenance jobs in hours/total hours of maintenance] × 100	<12
Preventive maintenance cost	[Preventive maintenance cost/total maintenance cost] × 100	20–30
Predictive maintenance cost	[Predictive maintenance cost/total maintenance cost] × 100	10–20
Annual maintenance cost	[Total maintenance cost/total manufacturing cost] × 100	<10
Maintenance revamping cost	[Revamping of asset/equipment cost/total maintenance cost] × 100	<5 (aging of plant is a criteria)
Downtime of plant/ equipment due to maintenance	[Number of times failed due to maintenance/ number of plant or equipment shutdowns] × 100	It must be a zero target <10
Monthly maintenance rework	[Job orders reworked/total job orders] × 100	It must be a zero target <3
Training coverage of maintenance team	At least on technical training year per employee	>95
Number of accidents due to maintenance work	Number of accidents due to maintenance work in a year	Must be a zero target
Energy consumption	Energy used in equipment	As per OEM recommendation
Safety hours	Total maintenance hours due to reportable accidents	Must be zero

lesser than the recommended target range (Table 4.2), the reasons for such are to be analyzed, and the corrective actions for achieving the targeted benchmarking range are to be analyzed. Accordingly, the necessary improvements must be incorporated in the asset maintenance strategic plan in operations-intensive organizations. Therefore, benchmark metrics are valuable measures in identifying the areas that are potential asset maintenance processes that are used for benchmarking of asset maintenance management. Deliberations on a few of the benchmark metrics mentioned in Table 4.2 are detailed below.

The *equipment/plant availability* metrics measurement has few challenges. One such challenge is how to measure the equipment running time. The equipment/plant may not be required to run due to less production capacity due to market conditions. In the equipment/plant availability metrics, *total running hours* is taken for calculation in the equipment/plant availability metrics, which means the equipment/plant non-production/idle time is excluded. In the case of standby equipment strategy or multiple production lines, clear definitions of metrics must be made to arrive at meaningful benchmark metrics.

In the *preventive maintenance schedule completion* metrics, the following needs to be considered: preventive maintenance is not just an inspection of equipment. After inspection of equipment, if any rectification of parts is required, it must perform such repairs. *PM scheduled hours* must built in such repairing time also in the *total scheduled hours.* In some cases, maintenance managers can decide to postpone such repairs depending on the severity of the problem to maximize the equipment's availability on a case-by-case basis and reduce the overall preventive maintenance cost. Therefore, operations-intensive organizations must develop appropriate guidelines for preventive maintenance in the maintenance policy for improving the accuracy of benchmark metrics.

The *annual maintenance cost* is calculated based on the *total manufacturing/ production cost*, and it must be less than 10% in case all asset maintenance activities are meticulously planned, scheduled, and executed timely and effectively. The organization's plan proactively has a cost advantage over the organizations that perform asset maintenance in reactive mode. These metrics are important to understand the efficiency of maintenance performance. In case annual maintenance cost is more than 10%, merely cost-saving measures on spare procurement or reduction in the workforce may not help the organization achieve the benchmark target of less than 10%. The necessary corrective actions are to be incorporated in planning and scheduling asset maintenance activities to achieve zero breakdown maintenance targets. The trade-off between breakdown maintenance and maintenance cost is required with proactive maintenance practices in operations-intensive organizations. Total maintenance cost majorly consists of workforce cost and material (spares) cost. The ratio between these two costs could provide a detailed analysis among the organizations for comparison in the benchmark study.

4.2.3 Challenges in Benchmarking

It is worth mentioning here that the benchmark metrics and key performance indicators (discussed in Chap. 2) are for different purposes in asset maintenance management. The key performance indicators are used to assess asset maintenance activities, i.e., internal to the organization. However, benchmark metrics are developed based on external comparison with the best player or leader in the industry to achieve business excellence and compete with its best players.

Benchmarking is not a one-time exercise, and it must be conducted every 3 years to review the benchmarking metrics developed for the organization. As shown in the benchmarking framework (Fig. 4.3), it is a cyclic process and a continuous improvement exercise for achieving asset maintenance excellence in the organization. Over the period, an organization can achieve a leadership position in the industry by adapting appropriate improvement steps.

The plant or process under benchmarking must have reached the matured state, i.e., implementing a systematic asset maintenance strategic plan after commissioning the plant or process. At least 2–3 years of asset maintenance performance datasets should be available in the organization to collate and develop benchmark metrics. If benchmarking exercise is done in the early stage of the plant, the expected business results could not be achieved due to the premature application in the asset maintenance processes.

In today's time, benchmarking becomes fancy in many organizations. Using benchmarking as a buzzword and showcasing the organizational capability to the stakeholders are becoming the industry practices. The real implementation of benchmarking does not happen in the organization, and even if they try to adopt few benchmarking practices from the peer organizations, they could not succeed due to the capabilities to drive the organization's results, which are limited.

4.2.4 Benchmark Study

The organizations usually hire consultants for conducting the benchmarking study. While conducting such benchmarking study in operations-intensive organizations, the following key steps are to be considered for ensuring successful completion of the study:

- Identification and finalization of plant assets/equipment-related systems and subsystems must be considered for the benchmarking study.
- Finalization of the consultant's scope of work to conduct benchmarking study related to all identified systems and subsystems.
- The next step is floating of tender and selection of the consultant for benchmarking study.

- Conduction of kickoff meeting to start the benchmarking project wherein the consultant representatives, the asset maintenance managers from the organizations, and other stakeholders are the meeting participants.
- Data collection by the consultant, i.e., the consultant shares data templates (physical and design data, financial and workforce data) to the organization under study. The organization's engineer-in-charge of the benchmarking project needs to provide necessary data in consultation with the respective stakeholders.
- The consultant does the data analysis and compares the data with global and country-specific peer organizations by performing a comparative performance analysis (CPA). The CPA is to be done on a quartile basis against various benchmark metrics (e.g., in natural gas pipeline systems, the calculation is done based on Equivalent Pipeline Complexity—EPC).
- Typical benchmark metrics considered for comparison are process and reliability, personnel, major maintenance, energy, and operating/maintenance expenses.
- The consultant submits the benchmark study report to improve asset maintenance processes for achieving the best targets compared to the industry leader.
- The report by the organization's asset maintenance managers underwent benchmark study and feedback to the consultant for any clarifications.
- The consultant submits the final report with an action plan for the necessary implementation.

In summary, the benchmark framework and its metrics provide the foundation for building benchmark study specifications before hiring consultants for conducting benchmark study or doing benchmark study internally by the organizations. Before conducting such a study, all the concerned stakeholders in the organization must be aware of the benchmarking process. It is suggested to identify at least five organizations in the industry for peer comparisons and find potential areas for improving the plant/facility under study. Even though benchmarking has prospective benefits, implementing the benchmark study's findings is crucial for reaping such benefits.

In this chapter, asset maintenance strategy and benchmarking frameworks with their processes are described holistically with the help of findings from the research study conducted among operations-intensive organizations. These frameworks provide guidelines for asset maintenance managers in the industry to develop asset maintenance strategic plans and implement best practices at an organizational level. The next chapter discusses on Intelligent Predictive Maintenance framework in the context of Industry 4.0.

Key Takeaways

- Standardization does not mean that all production plants/facilities must do asset maintenance strategies and practices in the same way. Specific asset maintenance strategies and practices are always essential, but standardization helps maintain consistency in asset maintenance practices and datasets across the industry's plants/facilities.
- To achieve standardizations in asset maintenance, organizations need to focus on the following significant areas at an organizational level. The areas are top management and organization structures, critical datasets, asset maintenance

practices, Information Technology, and asset performance management. Internal and industry norms guide the standardization at an organizational level.

- Successful implementation of strategies does not happen just like that, and it requires much effort by top management and the asset maintenance managers. Incorporating a structured approach allows organizations to connect all the asset maintenance processes and eliminate silos among the departments.
- The tools and techniques used in operations-intensive organizations are inadequate for deciding on the suitable asset maintenance strategy selection, formulation, and implementation. Therefore, it is essential and needs the current business scenario to upgrade predictive asset maintenance tools for asset maintenance decisions.
- EAM/CMMS system provides information regarding all the assets and ensures efficient utilization of assets/equipment over their life cycle, and finally, the system helps optimize asset availability and reduce costlier breakdowns.
- In the case of integration with the ERP system, organizations can determine how asset management impacts their financial objectives. An integrated system can handle depreciation and capitalization of the assets/equipment comprehensively in the ERP system by aligning with asset maintenance activities. Such integration helps ensure asset availability during its life cycle and maximize Return on Asset (ROA).
- It is essential to devise a set of benchmark parameters against a single parameter like asset maintenance cost. The operations-intensive organizations benefit from such measures in achieving cost saving, which is crucial in the current VUCA business scenario.
- The maintenance tactics other than preventive and predictive maintenance are RCM and TPM, which provide advantages to operations-intensive organizations to achieve asset maintenance excellence.
- *Proactive maintenance* is another maintenance tactic that uses asset maintenance practitioners' experience to decide on maintenance interventions in the plant/facility. The elements that affect proactive maintenance are maintenance personnel experience, limitations in operations, geographical area limitations, maintenance culture of the maintenance crew, and support from top management in implementing chosen maintenance strategy to practice benchmarked maintenance parameters.
- *Shutdown maintenance* is also called a turnaround shutdown. It is a planned shutdown of production/operations. Planned maintenance activities are carried out during the shutdown period (annually or every 2–3 years), requiring resource planning such as workforce, material, and other related requirements.
- Spares and inventory costs of asset maintenance account for 25%–45% of the total maintenance budget. Optimizing these costs is a challenge for maintenance practitioners since they are mostly involved in their major maintenance works, work order management, and preventive maintenance schedules.
- Intelligent diagnostic systems help predict equipment trouble and carry out efficient maintenance practices, resulting in maintenance excellence, operational excellence, and enhanced asset life cycle.

- The organization's maintenance policies in the industry aim for maintenance excellence, i.e., optimization of asset life cycle cost, replacement of assets in case of safety and environment regulations, and obsolescence of equipment/asset.
- Long-term financial budgeting must consider the asset's complete life cycle under planning, prioritization, and assessment of capital projects and their operating expenditures.
- Despite implementing the best asset maintenance strategy, unexpected asset/equipment failure is a ubiquitous reality. Therefore, maintenance managers must be ready to manage such kind of contingency management whenever it happens.
- Standardization of asset maintenance strategy processes is a challenge for maintenance managers in operations-intensive organizations since most organizations have multiple plants/facilities across the country or the countries.
- The *best-practice benchmarking* compares asset maintenance practices between the industry's indisputable leaders. The indisputable leader may be from similar industries across global locations. This type of approach could provide breakthrough improvements for the organization.
- Benchmarking framework is a cyclic process, and it is a continuous improvement exercise for achieving asset maintenance excellence in the organization. Over the period, an organization can achieve a leadership position in the industry by adapting appropriate improvement steps.

References

Komonen, K., Kunttu, S., & Ahonen, T. (2011). In search of the best practices in maintenance-new methods and research results. In *1st international maintenance world congress* (pp. 166–177). KP Media Oy.

Valkokari, P., Ahonen, T., Kunttu, S., & Horn, S. (2016). Fleet service solutions for optimal impact (in Finnish). Promaint—kunnossapidon erikoislehti. *Kunnossapitoyhdistys Promaintry, 30*(1), 36–39.

Velmurugan, R. S. (2016). Doctoral Thesis. In *Maintenance strategy and practices at a large gas utility company in India—A case study approach*. University of Petroleum and Energy Studies.

Wireman, T. (2010). *The maintenance strategies series* (5 Vols.). Reliabilityweb.com.

Chapter 5
Intelligent Predictive Maintenance: Industry 4.0

Abstract This chapter describes the relevance of asset maintenance management in the context of Industry 4.0. One of the essential digital strategies in asset maintenance is Intelligent Predictive Maintenance. A framework for Intelligent Predictive Maintenance based on the research study conducted by asset maintenance practitioners of operations-intensive organizations and the key elements of Intelligent Predictive Maintenance within an Industry 4.0 is presented. The key elements are the Internet of Services, Internet of Things, Smart Solutions, Smart Factory and cloud computing, and intelligent automation systems, which are discussed in detail in this chapter. Intelligent automation systems play a crucial role in implementing asset maintenance digital strategy at an organizational level.

Operations-intensive organizations in the industry have also started using the basic principles of IoT and Industry 4.0 to implement the related equipment and systems to form an intelligent network spanning its entire value chain. This type of self-controlled network system helps to identify the unexpected changes in the production process, predict equipment failures, and also prompt timely maintenance actions using the Intelligent Predictive Maintenance implementation cycle, which are described in this chapter.

5.1 Industry 4.0 and Asset Maintenance

It is imperative in current volatile market situations that the continual improvement in asset maintenance management for most of the operations-intensive organizations in the industry is inevitable. A new industrial revolution that has become apparent in the past few decades is known as Industry 4.0. The development of more extensive computer networks and their worldwide connectivity with real-time data access through Cyber-Physical Systems (Industrie4.0_Final_Report, 2013) are the key elements of Industry 4.0.

Industry 4.0 offers a comprehensive solution to all industries using the Industrial Internet of Things (IIoT) technology. It integrates physical and digital systems for better collaboration and connectivity among all the business stakeholders. Therefore,

R. S. Velmurugan, T. Dhingra, *Asset Maintenance Management in Industry*, International Series in Operations Research & Management Science 310, https://doi.org/10.1007/978-3-030-74154-9_5

Industry 4.0 empowers the business leaders to better control operations and maintenance of plant/facilities by driving growth through quick access of all organization datasets.

The implementation of Industry 4.0 concepts in the asset maintenance function domain can pose both opportunities and threats. The ways of doing predictive maintenance, monitoring of assets from remote locations, and managing critical assets change with digital connectivity of the assets within the plant/facility. A profitable asset maintenance management is a key to any organization's growth, as the case is with operations-intensive organizations. The organizations are continually improving their asset maintenance practices keeping face with fast-changing technology growth in the industry.

In the Industry 4.0 environment, asset maintenance and the overall optimization of the assets face many challenges. This environment is also offering many advantages. The new developments such as the Industrial Internet of Things (IIoT), big data, and advanced data analytics are proficient for maximizing asset utilization in its entire asset life cycle period. The advantages of such developments are increasing in asset availability, maximization of Return on Assets (Rona), and sustainability improvements.

Internet of Things (IoT) is a network of physical systems bearing the IP addresses for providing Internet connectivity and ensures digital communications among these systems. Industrial Internet of Things (IIoT) is a subsystem of IoT used for application in the industry to interconnect assets/equipment and provide data communications for managing such assets. IoT refers in the context of Industry 4.0 to IIoT. The datasets acquired over the period through IIoT decide to improve the asset/equipment's efficiency, increasing operational efficiencies. The industry estimation is that the organizations to the tune of 70%–80% may implement the IIoT in the next 5–7 years. Such IIoT implementation can result in 2%–3% of additional revenue for the organizations from the business operations (Koch et al., 2014).

The Industrial Internet of Things (IIoT) consists of hardware and software systems such as sensors, equipment, controllers, control system, data analytics algorithms, and datasets. The interconnectivity of hardware systems can be from a sensor, to a controller, to a control system. The data flow through this connectivity is accomplished with a physical network among the sensor, controller, and control system. The data analysis is carried out in the centralized server, and the output is integrated into Enterprise Asset Maintenance (EAM) system for quick action related to asset maintenance activities. The application of IIoT improves asset maintenance managers' decision-making, incorporates suitable changes in asset maintenance policy, and improves operational efficiencies (Zorzi et al., 2010).

Even after applying Computerized Maintenance Management Systems in operations-intensive organizations, the technology is not entirely utilized for various reasons. The reasons are asset maintenance is not fulfilled, design problems in the system, maintenance of sensors and their frequent failures, and the maintenance team does not possess the required skills to handle computerized automation systems. Intelligent Predictive Maintenance (IPM) can come to the rescue to improve leveraging technology by optimizing the maintenance activities. The IPM uses the conventional condition monitoring methods embedded with the IIoT technology to monitor equipment/assets. This kind of technology provides smart choices to achieve maintenance excellence in the organization and minimize asset maintenance costs (Schaeffler Technologies AG & Co., 2015).

The organizations have started using the IIoT technology and Industry 4.0 guidelines for interconnecting the assets/equipment and systems through a smart network within the business portfolio across the value chain. Such an asset maintenance-controlled network supports the Intelligent Predictive Maintenance (IPM) framework to predict the equipment failures in advance, understand various changes in the process operations, and indicate the required asset maintenance jobs for special equipment. The IPM framework with the digital technology of Industry 4.0 aims for near-zero downtime of equipment/plant in operations-intensive organizations. Due to the industry's complexity, the standard IPM framework does not fit all organizations. The framework's customization is based on the industry's type, the complexity of the process/operations, and maintenance objectives.

The industry's IPM framework must achieve the following business objectives: increasing asset/equipment maintenance reliability, optimizing operations toward business excellence, and creating additional value across the organization's business vertical. Therefore, organizations in the industry should place themselves to obtain the maximum out of asset maintenance using the IPM framework in the environment of Industry 4.0.

Predictive maintenance deals with applying IT infrastructure with custom software applications of the past's specific assets/equipment. In case of any upgrade or modification in asset/equipment, the new software application must integrate into the assets/equipment. Such integration causes inconvenience to the asset maintenance engineers, and they simply went back to the conventional way of doing the asset maintenance. With the development of IIoT digital platforms, predictive maintenance applications can be developed and supported through the cloud's IT infrastructure. The digital platform manages the connected assets/equipment, datasets management, and user management. It is just a requirement to reconfigure the asset maintenance services and related sub-services to modify or upgrade the asset/equipment. Such systems focus on creating predictive algorithms to perform specific asset maintenance activities and add value to entire asset maintenance management.

Box 5.1: Advantages of IIoT-Enabled Predictive Maintenance

- Prevention of unplanned downtime.
- MTBF approaches infinity.
- MTTR goes to optimum level.
- Perform maintenance when required.
- Operations downtime avoided due to reschedule of the production time.
- Costly emergency repairs can be avoided.

IIoT-enabled predictive maintenance provides many advantages to operations-intensive organizations to prevent or delay asset failure. The list is as shown in Box 5.1. Before any asset failure, the asset availability time depends on the historical data and the associated algorithms developed in the system. The system provides alerts before asset failures, driven by the operations that allow the maintenance team to schedule their maintenance jobs according to the system alerts. Therefore, a well-developed predictive maintenance system with IIoT enabling would avoid nearly all production losses.

5.2 Intelligent Automation Systems

Digital transformation of operations-intensive organizations mainly focuses on the questions such as what technologies are available for the industry and how such technology can improve its business processes. The evolving technologies such as Robotic Process Automation, the Industrial Internet of Things, and Artificial Intelligence are the building blocks of intelligent maintenance systems. Many organizations are still exploring the utilization of these systems in maintenance processes. Integrating digital technologies with human potential, managing the integration process, and incorporating new digital tools in asset maintenance management are crucial for any business in the industry. IoT is not just the deployment of digital tools in intelligent automation; it is a new way of performing the jobs by maintenance managers/engineers that allows the organizations to continually understand, adjust, and develop as they attempt to achieve the corporate objectives. Therefore, an intelligent automation system is an innovative blend of competent workforce and digital workers to gain the complete advantage of promising technologies.

Intelligent automation system (IAS) brings together modern technologies in asset maintenance such as portable, analytics-based, cloud computing and intelligent-driven automation. IAS further integrates these components with the skilled workforce to produce value across the business. It is not only a software application and involves a unique way of philosophy because it is a functional strategy at an organizational level. There are many benefits for an organization if intelligent automation is introduced in asset maintenance. Intelligent automation systems' major components are data scientists, cognitive automation, intelligent worker, and intelligent communication tools. These components are shown in Fig. 5.1.

Artificial Intelligence and neural networks create massive data and help operations-intensive organizations to make decisions faster and more accurately than human decisions in asset maintenance management.

Data scientists in the organization analyze the big data available and work upon questions such as the following: Which part of the asset/equipment provides the signals of wear and tear, and is there a requirement to repair that part or replacement? What type of maintenance tactic can be used for the specific asset/equipment? The deployment of digital technologies enables the maintenance team to forecast the failures and respond quickly to changing conditions.

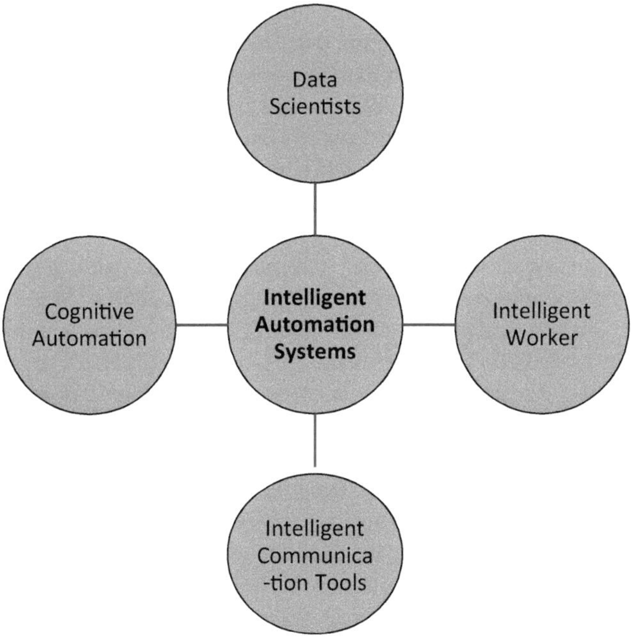

Fig. 5.1 Components of intelligent automation systems

Intelligent communication tools are RFID-based communication, chatbots, and intelligent automation systems. The maintenance technician/operator who works in the plant/facility requires observing various process parameters such as vibration, pressure, temperature, and location of leaks at the site and logs those details manually in the logbook, which are in the computer-based system. The intelligent communication tools help the technician/operator to speak to the smart device. The device automatically detects the asset location, type, equipment details, and specifications. The chatbot-like application communicates the site observations to Robotic Process Automation (RPA) software or any Enterprise Resource Planning software to create job orders or logbooks. Such automation aids in avoiding the manual entry error of operator/technician and time delay in reading the logbook and then initiating a job order by the operations engineer.

Cognitive automation uses tools such as machine learning, variable format processing, and keyword-based recognition. Cognitive automation in asset maintenance captures and uploads data in Enterprise Asset Maintenance (EAM) or Enterprise Resource Planning (ERP) systems for use across the organization. For example, usually, materials store employees write-down and enter data manually in the computer system for any new spare part of equipment receipt and inspection for individual shipment. In intelligent automation systems, the cognitive automation tools help capture the packing slips' image and automatically update the EAM/ERP system's inventory data.

Intelligent workers use intelligent automation tools and cognitive automation for processing plant shop floor information, datasets, and automated emails by integrating the tasks and job orders of asset maintenance of the plant/facility. The intelligent worker can be digital labor or digital assistants, i.e., machine workers. Such work-based robotic machines help human workers execute highly repetitive and volume jobs with greater efficiency and as quickly as possible. Therefore, an organization can benefit from intelligent/smart workers to maximize maintenance efficiency and minimize maintenance costs.

Operations-intensive organizations traditionally deal with the Information Technology-driven approach in implementing the digitization of asset maintenance systems, i.e., choose a reputed vendor who is having a proven track record in digitization in asset maintenance management, procure the software applications, and roll out across the organization. It is not merely adopting software applications, and organizations need to deal with intelligent automation system implementation as a strategic issue. The top leaders of the organization must think holistically in integrating intelligent automation systems by considering the questions such as what are the long-term/short-term business objectives, how IAS helps achieve the corporate results, and what competencies are required to accomplish the digital vision. The answers to these questions are to be incorporated in digital strategy in designing the components mentioned above of intelligent automation systems to ensure their value delivery and derive maximum benefit for the organization in operations-intensive organizations. For example, Industrial Internet of Things (IIoT) can capture many asset maintenance datasets from the equipment/instrument. Such data has no importance if the organizations cannot evaluate and make decisions using intelligent automation tools.

5.3 Intelligent Predictive Maintenance—Key Elements

The key elements contributing to Intelligent Predictive Maintenance (IPM) in the Industry 4.0 context are the Internet of Services, Industrial Internet of Things, cloud computing, Smart Solutions and Smart Factory (Velmurugan, 2019). These elements are depicted in Fig. 5.2.

Internet of Services (IoS) is the set of datasets responsible for ensuring communication among the network of computers, i.e., the Internet. The subsystems classified under IoS are *Service-Oriented Architecture (SOA) and Web Service Interface (WSI)* for technology and business communications. These services aim at connecting servers, systems, and subsystems for providing access to people. The system software integration with various end systems (mobile devices) enables mutual communication, any time connectivity, and system support by properly deploying the system processes. The services integration must ensure secure, safe, and reliable communications among the devices and systems. The IoS sub-services are shown in Fig. 5.3 (Schroth & Janner, 2007).

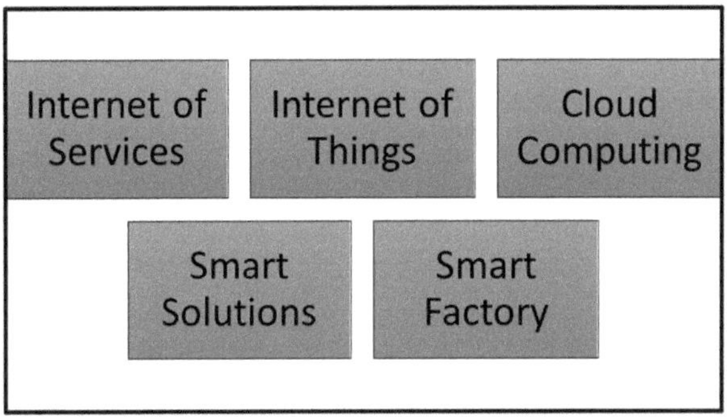

Fig. 5.2 IPM: key elements

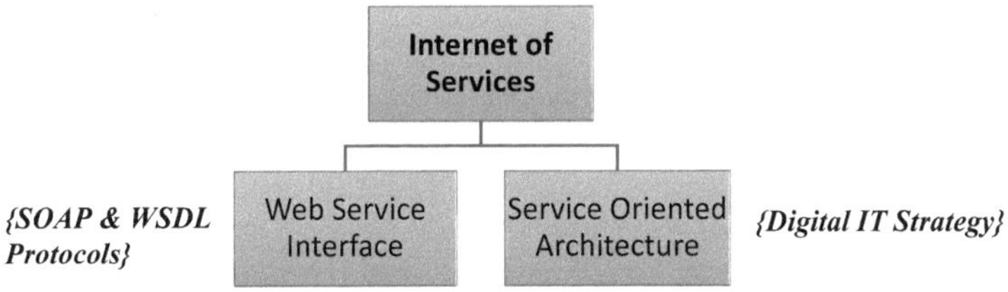

Fig. 5.3 Internet of Services

In Web Service Interface (WSI), Simple Object Access Protocol (SOAP) over Hypertext Transfer Protocol (HTTP) allows third-party applications to call functions and, further, to read inputs and outputs using Web Service Definition Language (WSDL) based on the configuration of any automatically generated transactions.

One example of WSI in operations-intensive organizations is integrating the Supervisory Control and Data Acquisition (SCADA) system and Enterprise Asset Management (EAM) system. The real-time access of an asset/equipment can be carried out through Web Service Interface between the systems over the protocols mentioned above. Another example is Programmable Logic Controllers used in emergency shutdown application of process plants/facilities through a web browser. Therefore, such web services make it possible to obtain information from various asset maintenance systems, locations, and networks and be available to asset end user without any additional efforts to get such datasets or information. Further, all asset maintenance systems can be integrated with an Enterprise Resource Planning (ERP) system for a company-wide integration.

Service-Oriented Architecture (SOA) is not a protocol, and it is a digital strategy for connecting the resource across the organization. Integrating the distributed resources helps the asset maintenance managers identify the plants/facilities' standard capabilities and effectively leverage their utilization. SOA can be used in asset maintenance function as an Information Technology strategy to integrate the various asset maintenance systems operating in silos to drive faster implementation and reduce maintenance cost by making available resources of distributed systems and assets in standard methods. The basic principles of the SOA strategy are as follows:

- Service interface and resource implementation are to be treated separately. That means the clients use the services that are not concerned with the implementation of the services.
- The service can be identified dynamically over the network.
- The available services are self-controlled to perform the predetermined tasks and ensure their independence; such services are freely coupled.
- Combined services can be developed by aggregating the available services from the set of resources.

In a SOA-integrated asset maintenance system, all resources are registered in their services in the hybrid system, and an asset maintenance end user uses the registry to find the services essential to perform the current asset maintenance task. Web services are a standard way to execute a SOA strategy. The well-thought application of SOA digital strategy impacts the asset maintenance implementation processes in operations-intensive organizations in the coming years and provides much comfort to maintenance managers/engineers.

Internet of Things (IoT) connects many sensing devices across the counties worldwide through Internet protocol version 6 (IPv6) (Indsutrie4.0_Report, 2013). Such network (Internet) connectivity can link more than one billion people through social network platforms and connect asset maintenance systems across the organization's plant/facility over the network. This kind of connectivity is possible to achieve duc to less expensive and innovative sensing devices. IoT is a *World Wide Web* that interconnects the network of computers, systems, and subsystems across the universe for the free flow of information (Yan et al., 2008). These systems are identified using an Internet Protocol (IP) address unique to the computer, systems, or any subsystems connected over the network. IoT is a subsystem of the Cyber-Physical Systems (CPS). In the case of *Industrial IoT* (*IIoT*), the connected systems can be field instruments, equipment, servers, and control systems that are part of its asset maintenance network. IIoT can be considered a subset of the Cyber-Physical Systems and can also be part of an organization's intranet.

The devices/sensors of the physical assets/equipment and Information Technology systems communicate and exchange the data over the Internet using the Internet of Things. A conventional CMMS software integrated with the IoT automatically communicates among the connected devices/sensors/control systems without human intervention. Such connectivity eliminates human errors, saves workforce use, and has quick availability of data over the network. For example, equipment like gas turbine consists of several sensors/devices connected to this. The data from these sensors can be automatically updated in the gas turbine control system, and all these data are also automatically updated using IoT in other connected systems such as Manufacturing Execution System and Enterprise Resource Planning system. Therefore, gas turbine's healthiness parameters like running status, vibration, pressure, and temperature are available to the entire asset maintenance team. Asset maintenance manager can take proactive action if the gas turbine requires immediate maintenance intervention, and a job order can also be created automatically. This kind of intervention helps the maintenance team avoid downtime of equipment and facilitate the operations team to achieve the production targets or maximize the production.

Box 5.2: Advantages of IoT in Asset Maintenance Management

- Decrease in asset maintenance cost.
- Increase asset availability time in between failures of an asset.
- Increase ROI by maximizing the production.
- Ensure sustainability at an organizational level.

The Internet of Things can automate the asset maintenance process to collect the relevant information using field sensors attached to the asset/equipment and update the central asset maintenance systems. Organizations depend on the operators to sense the failure from available data and inform maintenance engineers of any corrective action. IoT technology eliminates the requirement of operator-based maintenance. If any sensor failure occurs, the information is sent to the central asset maintenance system and simultaneously informs field maintenance engineer. The field engineer/technician immediately corrects the failure, and asset downtime can be avoided due to immediate attention. IoT technology helps to track equipment/asset performance data from designated sensors of the equipment. An asset maintenance manager can predict the failure and assign the field engineer/technician to investigate before intimation from the operation engineer/operator from the field from these performance parameters. To reduce maintenance costs, the asset maintenance manager needs to identify the source from where the significant contribution emerges out. Therefore, identifying the root cause of any failure in asset maintenance is crucial to fixing the failure and avoiding the same failure. IoT-enabled asset maintenance system supports identifying the root cause of the failure and makes smart financial decisions. The advantages of integrating IoT technology with the existing asset maintenance management systems are shown in Box 5.2.

The operations-intensive organizations are much interested in continuous tracking of the assets/equipment deployed in the organizations to maximize the Return on Assets. Currently, most organizations have been using IT-enabled systems such as SCADA-, EAM-, ERP-, and RFID-based solutions. All the systems are coexisting and being used by various stakeholders for different purposes within the organization. These systems are distributed across the departments, work centers, plants, facilities, and countries, and they are not integrated for leveraging the standard processes among the systems' distribution. IoT-enabled smart asset maintenance systems integrate all these standalone systems and create an integrated system to contribute as a unified strategic system for business optimization and disrupt the business processes. Such a single asset maintenance system continues to perform all the activities such as monitoring asset condition, managing asset life cycle, and controlling asset maintenance processes in the existing asset maintenance management system. It also enables workflow automation, indications on a real-time basis, automated data analysis and predictions, dynamic control of asset processes, and

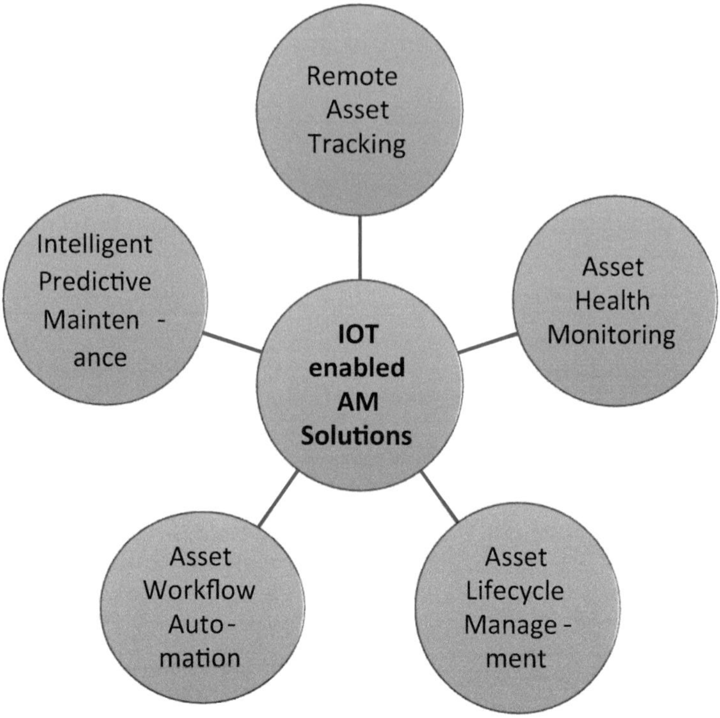

Fig. 5.4 IoT-enabled asset maintenance solutions

smart predictive maintenance solutions and provides real-time datasets with cross-functional data analysis and business information. The IoT-enabled asset maintenance solutions are depicted in Fig. 5.4.

The solutions shown in Fig. 5.4 are a few of them, and there can be many other solutions based on the application in the industries. This chapter discusses the Intelligent Predictive Maintenance framework based on the operations-intensive organizations' research study.

Cloud computing is a method to share computing systems' processing resources among the users who require computing services. These services can be availed at any time, on-demand basis, and over the network across the universe. Cloud computing is a practical model to use the networks, computing servers, storage, services, and various applications from the shared virtual servers, and this is configurable with user requirements (Mell & Grance, 2011). Industry 4.0 guidelines enable the industry to use cloud computing services for industrial applications (Bai et al., 2012).

These applications also include asset maintenance management in organizations. It is possible to quickly specify and deploy the asset maintenance related to cloud computing services with fewer efforts by the asset maintenance team in operations-intensive organizations. Cloud computing can be classified as attributes, service

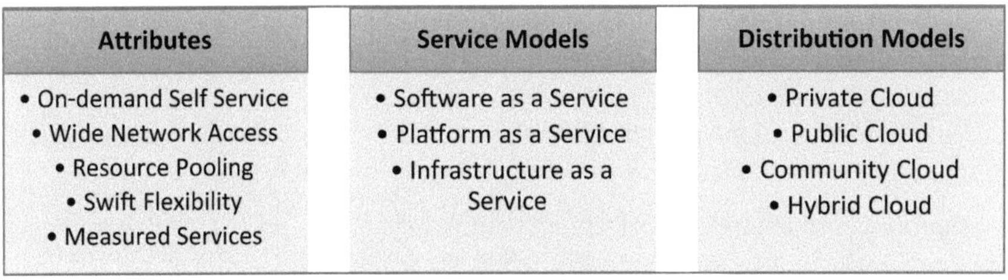

Fig. 5.5 Cloud computing elements

models, and distribution models, as depicted in Fig. 5.5. The Intelligent Predictive Maintenance systems can utilize cloud computing services to share a tremendous amount of asset maintenance data within the organization's asset maintenance network (Velmurugan, 2019).

Organizations can hire computing resources like servers (physical/virtual), storage, and networks for industrial applications in the infrastructure-as-a-service model. The organization required to build asset maintenance applications can use such online infrastructure, and it helps them quickly innovate and reduce the cost of deploying new applications. The industry applications may require a platform and software tools for developing the application in asset maintenance management. The organization may then opt for a *platform-as-a-service* (*PasS*) service model that provides operating systems (OS), application software, management of the database, and development modules. Further, if the organization can avail some asset maintenance software applications that required direct access of software over the cloud, then the software as a service (SaaS)-based solution from Original Equipment Manufacturers is the option. In this service, the asset end user needs not to be concerned about hardware service or software service as in the cases mentioned above. The user can log on to the services directly using any standard application through a web browser.

The *cloud computing attributes* are described as follows:

- The resources like network storage and server time allocated to the customer can be accessed on-demand basis. The data access happens instantly without any interruptions automatically. This kind of service is known as *on-demand self-service.*
- The various customer devices (laptops, computers, tablets, mobile phones) can obtain the right to use over the Internet and access using standard processes from the shared cloud computing resources. This attribute of cloud computing is *wide network access* capability.
- Cloud computing resources are shared among several customers worldwide by pooling such resources through a dynamic allocation process. This capability of cloud computing attribute is *resource pooling.*
- Resource pooling of processing resources such as the network's bandwidth, storage, and memory in cloud computing is combined to serve several consumers worldwide by dynamic allocation process.

- *Swift flexibility* capabilities are to manage the changing demands in computing requirements, and if it requires, the resources can be appropriated by the cloud computing vendor.
- *Measured services* have control over the resource allocation to the consumers and report to ensure transparency between providers and consumers.

The above characteristics of cloud computing are leveraged, and computing resources must be used appropriately while designing asset maintenance applications for plants/facilities. The private cloud runs the cloud service on the campus to ensure complete control over the requirements of critical workloads and confidential data compliance. The public cloud is paying as service providers, and the consumer can quickly get access, test applications, and launch the processes as quickly as possible. A hybrid cloud is nothing but using these two types of deployment to the extent possible based on the individual consumer or organization's requirements. On the other hand, a multi-cloud deployment model uses the computing resources from more than one cloud to optimize the cost, service availability, and geolocations. The choice of cloud deployment model in asset maintenance system applications depends on the organization's types of assets, business requirements, and system integration choice.

Industry 4.0 concepts are crucial to profitable operations-intensive organizations in the future. Smart products and smart services are the two main factors contributing to **Smart Solutions**. The earlier discussed Cyber-Physical Systems could also be called **smart products**. CPS can build connectivity through modules such as equipment-to-equipment communication, embedded network interfaces, and human-machine interface (HMI). Network connectivity plays a crucial role in extending the new features and functions of these smart products. The advanced delivery models and service support of *smart services* provide innovative methods for an organization in the operations-intensive industry to create new business models and emerging markets in their business domains.

Smart products' capabilities combine with smart services offer Smart Solution, which paves the way for better self-management and improving communication facilities. Therefore, in the era of decentralized single business unit (SBU) models of decision-making and business operations, these Smart Solutions offer various effective business models for the industry (Lee & Wang, 2008).

The CPS utilization expanded to the manufacturing plant/facility environment in an asset maintenance setup called Cyber-Physical Production Systems (CPPS). These systems offer Smart Solutions in the industry environment, such as pre-warning analysis reports of the asset/equipment condition to mitigate unexpected failures. Such types of solutions aid predictive maintenance applications to make smart decisions to maximize the asset life cycle period. For example, gas compressor and its associated control systems can be equipped with CPS for self-maintaining its operations and maintenance and generate the maintenance trigger automatically, which alarms the operator to initiate the maintenance support. Further, the compressor's connectivity through smart services with the Original Equipment Manufacturer (OEM) of the compressor ensures the remote connectivity to generate the condition monitoring reports, planning and executing maintenance schedules, and optimizing the production.

Smart Factory comprises a set of assets/equipment networked together for self-maintaining its associated systems with the optimization layer to form a decentralized operations/production control. Apart from this physical networking of the systems, the datasets generated from the asset maintenance system of the particular asset/equipment are captured in the Smart Factory systems, supporting an innovative data-driven approach in the industry to achieve operational excellence (Lucke et al., 2008). Therefore, the Smart Factory concepts are the key components in Maintenance 4.0 with the broader ambit of Industry 4.0. Calm system (hardware) and the context-aware applications (software) are the significant elements of Smart Factory. These elements are used to build a physical network with the asset/equipment and apply the innovative computing systems to establish a remote physical system.

The virtual physical system must update the field datasets on a real-time basis from physical sensors embedded in the asset/equipment. The Smart Factory applications must develop error-free and robust systems to cope with field sensor disturbances. In case of failure, the built-in mechanisms of Smart Factory should facilitate quick recovery from the failures using self-correcting processes. *Smart algorithms* are used in cyberspace to capture the physical data from the field sensors on a real-time basis, ensure correct measurements, and collate and compute the equipment/asset data as per requirements to predict asset performance and health condition (Lee et al., 2014).

Smart Factory concept is applied to operations-intensive organizations where operations form a portion of smart systems, and all the asset/equipment is put together using machine-to-machine (M2M) communication or other smart algorithms. A well-designed wireless network with easily replaceable modules is critical in enhancing operational efficiency. The Smart Factory environment provides better reliability, availability, performance, and quality for improving the asset life cycle.

The system in which the various types of intelligent devices are interfaced and collaboratively communicated through a smart system with an environment of Smart Factory is known as Factories of the Future (FoF). This approach builds a pooled network of independently connected and decentralized smart devices. Factories of the Future networks are quickly configurable with a plug-and-play approach to meet the business objectives (Colombo & Karnouskos, 2009). FoF is established for improving the product specifications and its quality continually to ensure a customer-centric approach in all the business processes (Souza et al., 2015). In the operations-intensive organization's manufacturing environment, FoF's objectives are maximizing productions with zero defect products, improving asset reliability with energy efficiency, and increasing business profits with sustainability.

5.4 Intelligent Predictive Maintenance Framework

The key elements of Intelligent Predictive Maintenance (IPM) emerged from a literature study, and key enablers arrived from industry reports of operations-intensive organizations that conceptualize the IPM framework in the context of Industry 4.0. The framework is shown in Fig. 5.6. The IPM framework in Industry 4.0 supports the businesses to achieve operations and maintenance excellence by upgrading conventional systems to intelligent embedded systems with Smart Solutions, i.e., Internet of Things and the Internet of Services. The asset maintenance processes with the aim of production/capacity maximization supported through Smart Factory concepts such as cloud computing and Factories of the Future are optimized for accomplishing asset maintenance objectives.

Overall, Intelligent Preventive Maintenance framework comprises Smart Factory and Smart Solutions that facilitate innovative business models. The smart services in the industry are used to ensure reliability and integrity of assets/equipment and increase the scale of operations with the help of datasets generated through field sensors/devices embedded in the plant/asset/equipment.

Implementing cloud-based systems in the operations-intensive organization's asset maintenance processes is one example of smart service in the IPM framework. The above conceptual framework (Fig. 5.6) is also validated through empirical research carried out from asset maintenance managers and operations managers from operations-intensive organizations through focus group discussions (FGD) and semi-structured interviews using the case approach (Yin, 2003).

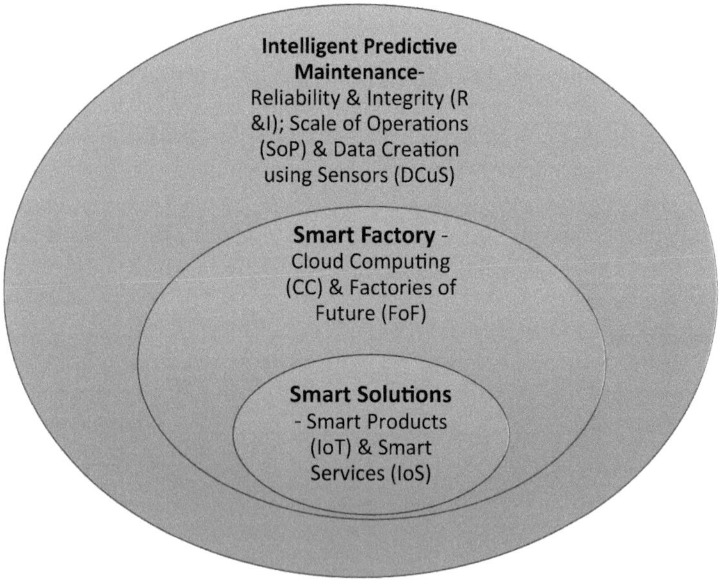

Fig. 5.6 Intelligent Predictive Maintenance framework

The empirical findings from the research work conducted in operations-intensive organizations are generalized with the outcomes based on the IPM framework, which provides guidelines to the asset maintenance managers for implementation of Intelligent Predictive Maintenance at an organizational level.

5.4.1 Smart Solutions Enablers

Internet of Things is a real technology enabler in providing Smart Solutions in operations-intensive organizations. In an industry environment, IoT is recognized as the Industrial Internet of Things (IIoT). IoT configures automation systems in the plant/facility that integrate the industrial legal protocols such as Ethernet, Modbus, Highway Addressable Remote Transducer (HART) protocol, and Simple Network Management Protocol (SNMP) and provides actionable real-time data and insights. The asset maintenance administrators can operate the flexible automation system with remote real-time monitoring, diagnostics, and rectification of symptoms before the failure happens. Such an automation system is also useful for operational planning, inventory management, and collecting feedback from all the business stakeholders. The connected automation control systems create diverse sets of devices, capabilities, and network protocols, causing complex operations and security challenges. Therefore, Smart Solutions are integrated with IIoT security, combined with identity management and encryption technologies to build the connected automation systems with trust at their roots. These solutions enable to address data, security identifications, and technology encryptions across the environments and devices through their system life cycle.

Even though the Industrial Internet of Things (IIoT) is much discussed in the Industry 4.0 context, connecting "Things" and providing them with intelligence is the real innovation, i.e., the Internet of Services. In operations-intensive organizations, recognizing the Internet of Services and exploiting the opportunity maximize their operations. For example, the individualization of mass production maximizes the revenue of the organization. A smart manufacturing facility/plant must be flexible enough to produce intelligent products, and such value addition in products could yield additional revenue in final products. Therefore, it is not a cost-saving measure, and it is a business model, i.e., providing value addition in the product, earning more revenue, and achieving profitability in the business. It is a cautious note to operations-intensive organizations that they are merely adding technology infrastructures like smart products and services, which cannot yield the result. A well-thought strategy by keeping the value addition of a product in mind allows justifying the organizations' digital investment.

5.4.2 Smart Factory Enablers

Factories of the Future in terms of Smart Factory describes the future of the operations/production using Smart Solutions and cloud computing. Such operations are intelligent, flexible, and dynamic. The operation processes are organized so that the complete chain of operations from the supply and logistics and the complete life cycle of the production are connected to business boundaries. The operation processes impacted are manufacturing planning, product expansion, logistics, Enterprise Resource Planning, Enterprise Asset Management, Manufacturing Execution Systems, control systems technologies, asset embedded devices, and field instruments. Smart Factory implementation provides the capabilities to improve asset maintenance processes through self-optimization and autonomous decision-making. It is not the current capability of the asset maintenance programs in operations-intensive organizations.

Today's automation solutions consist of four layers such as filed instruments/ devices (sensors, actuators) layer, control systems layer, process management (Distributed Control System, SCADA, and MES systems) layer, and enterprise management (ERP system) layer, as shown in Fig. 5.7.

In the Smart Factory, the field instrument layer remains the same in Traditional and Smart Factories as a dedicated independent level. The field sensors, actuators, and other devices are intelligent in the case of Smart Factories. These devices are connected to Cyber-Physical Systems and perform autonomously several processes, and additional hardware is fit into existing field instruments. The functions related to control systems, process management, and enterprise management are mapped to high-functioning servers established in a group, data hub, or cloud system.

Factories of the Future take care of virtualization, a partition of a specialized function, and the computing hardware. The advantages of Factories of the Future are that it minimizes the number of devices so that the resources can be managed easier,

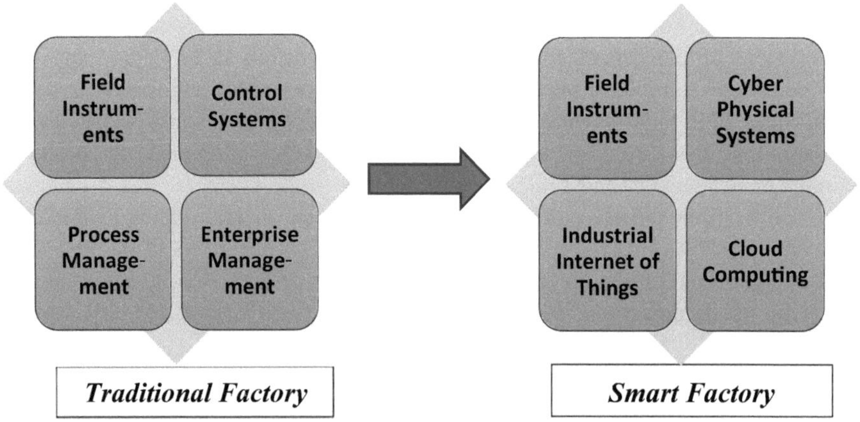

Fig. 5.7 Traditional Factory vs. Smart Factory

and cost savings can also be achieved. The FoF concept is not yet adopted in industrial automation by many organizations due to poor connectivity of the field instrument layer with the servers. All the filed instruments/devices must be upgraded as smart devices by upgrading the network bandwidth and speed. The other issues are performance, reliability, and determinism of the asset maintenance team.

5.4.3 IPM Growth Drivers

The assets/equipment are maintained by performing scheduled maintenance of the assets/equipment to replace spares and servicing the equipment. Such a maintenance philosophy increases the planned downtime of the equipment and shutdown of the plant operations. Further, the availability of spares is also critical in asset maintenance management. Unavailability of assets/equipment results in plant downtime. On the other hand, increasing spare inventory leads to holding the working capital and increases the risk of obsolescence that eats away the organization's bottom line. Therefore, the primary objective of Predictive Maintenance is to break these trade-offs of spare inventory by encouraging organizations to capitalize on the useful life of the equipment spare parts while preventing unplanned downtime and reducing the planned downtime.

With the introduction of Industry 4.0 for operations-intensive organizations, the organization can leverage technologies such as Smart Solutions and Smart Factory. Equipment-to-equipment interface (EEI) and human-machine interface (HMI), in collaboration with the diagnostic and cognitive technologies, are the key elements to facilitate quick decision-making and correct maintenance intervention in asset maintenance management. The datasets are captured from diverse Predictive Maintenance-based asset maintenance management systems such as ERP, EAM, and CMMS. Smart Solutions, combined with innovative prediction models and analytical tools, help asset maintenance managers forecast asset failures and proactively rectify them. The Smart Solution changes the predictive maintenance to Intelligent Predictive Maintenance. These intelligent systems adopt advanced machine learning technology to improve the precision algorithms' precision to maximize asset maintenance performance.

5.4.4 Role of Sensors in IPM

Asset maintenance managers need to choose the applications of Intelligent Predictive Maintenance in asset maintenance. Once it is done, the prime task is to identify the data sources and how they can be captured. For data from ERP/EAM/MES systems, there is a requirement of integration with Smart Factory systems with these systems. The other data need to be captured through smart devices/sensors, and this is an additional hardware requirement. The key to the successful implementation of the Intelligent Predictive Maintenance program is using the appropriate sensors/devices to determine the asset/equipment condition. Sensors/device selection is based on the equipment's predicted failure modes. For example, the failure modes of rotating machines include bearing failure, cavitation, anti-surge, unbalance, high vibration/temperature levels, and low lubrication system flow/cooling water flow.

The foundation of Industry 4.0 is collecting real-time data from these sensors and analyzing the data. Then, the input/output link flow such as data gathered to various Smart Factory systems and ensured to achieve all the aspects of Industry 4.0 becomes a reality. The data from the field sensors flow to cloud systems through field gateways and cloud gateways. The field gateways filter and preprocess the data, which are the physical systems. The cloud gateways' role is to ensure secure data transmission, network the field gateways through various protocols, and provide connectivity. The streaming data processor, which is part of the cloud systems, allows a quick and continuous data flow while ensuring efficient data transmission.

Further, the data flow goes to data storage; it stores the structured and cleansed data for further computing use. Machine learning (ML) algorithms are incorporated into the system to identify the correlations in datasets and discover unusual data patterns. The established data patterns are shown in asset maintenance predictive models. The predictive models are regularly updated and tested for accuracy in their outputs. If the model's output varies with the expected data, then the predictive model is revised, reprogrammed, and tested again. Such modeling is an iterative process until the desired outcome is achieved from the predictive maintenance models. It is understood from this predictive model building that data is critical information for the model development. If the initial data is wrongly captured or data with error leads to a predictive model's wrong development. In that case, Intelligent Predictive Maintenance does not yield the desired results.

5.4.5 IPM Implementation

As described above, the significant enablers of Smart Solutions and Smart Factory and the growth drivers of Intelligent Predictive Maintenance (IPM) with sensors help contribute to effective implementation of IPM in the operations-intensive organizations in the industry. An Intelligent Predictive Maintenance implementation cycle (D^2 IED—Develop, Deploy, Integrate, Execute, Document model) is developed through an empirical study [case approach (Yin, 2003)] conducted among operations-intensive organizations. This model is shown in Fig. 5.8 and provides directions to asset maintenance managers to implement Intelligent Predictive Maintenance systematically in the organization.

In conventional predictive maintenance systems, the sensors/devices embedded in the asset/equipment are not smart. The transformation from conventional predictive maintenance to Intelligent Predictive Maintenance using the IPM implementation model generates the additional datasets from newly *installed intelligent field sensors*, and such data generated are shared over the *asset maintenance networks*,

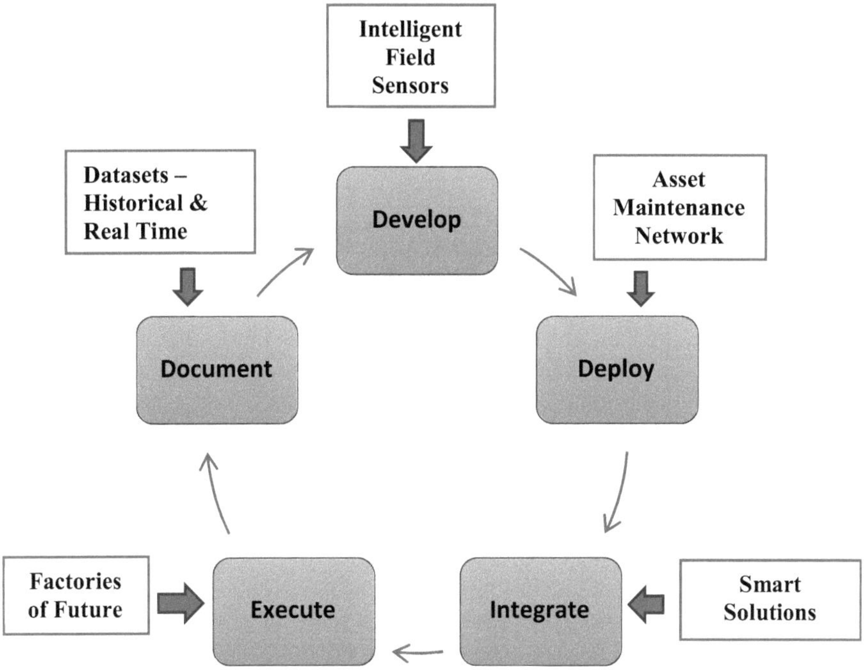

Fig. 5.8 Intelligent Predictive Maintenance implementation cycle

including Distributed Control System networks through various interfaces such as machine-to-machine, human-machine, and control system-to-control system. The IPM infrastructure establishes pooled communication among the industrial network systems and control through this network.

The vast information/datasets generated through IPM infrastructure are used centrally to collate, analyze, and infer data patterns and integrate with other datasets to apply asset maintenance decisions for effective implementation of the asset maintenance activities at the plant/facility level. These datasets are also managed based on the legal, regulatory, technical, and social international/national standards at an organizational level.

The *analytical models* incorporated in the Intelligent Predictive Maintenance systems using new technologies facilitate the initiation of the asset maintenance actions automatically and trigger the job order to the asset maintenance team to promptly perform the required asset maintenance activity in the asset/equipment to promptly avoid unexpected failure the asset/equipment. Therefore, *Smart Solutions* can be used in operations-intensive organizations to implement intelligent predictive asset maintenance better.

The implementation of intelligent automation systems in asset maintenance under Industry 4.0 is a value creation for asset maintenance managers to achieve the critical asset maintenance objectives such as increased asset reliability, maximized asset availability, and quick retrieval of the asset after the failure. The IPM implementation cycle's functional configuration confirms the availability of asset/equipment historical data and predictive maintenance information of all the assets/equipment across all the organization's facilities/plants through distributed cloud network universally. It can also be interfaced with the organization's intranet. The Factories of the Future components can be well integrated with the IPM implementation cycle during its execution phase.

Overall, the D^2 IED model enables to maximize the asset/equipment availability and reliability, reduces the human intervention in asset maintenance, and improves asset safety performance since the Original Equipment Manufacturer (OEM) can provide consultation in predictive maintenance decisions based on data generated through smart sensors/devices embedded in the asset/equipment from the remote location without any geographical boundaries using *digital twin* technology. Therefore, as explained in this model, asset maintenance datasets are developed, deployed, integrated, and executed to improve the efficiency of predictive maintenance systems and models at an organizational level. Though the IPM implementation cycle does not end here, the documentation of datasets and information are essential for further analysis, and historical datasets of asset/equipment can be used in the future for asset maintenance decisions in the organization.

IPM implementation is possible with robust technological infrastructure, which requires initial investment to add smart systems/sensors/devices. However, implementing such smart predictive maintenance systems provides huge asset maintenance cost benefits and ensures effective implementation of asset predictive maintenance programs at an organizational level in operations-intensive organizations. The integration of cloud computing with asset maintenance systems across the business vertical's value chain in the organization is a case of Industry 4.0.

The advantages of Intelligent Predictive Maintenance implementation are huge for asset maintenance managers. The remote access of IPM systems through networks is highly vulnerable to cyberattacks; sabotaging the business's assets is a concern. The well-organized cyberattack prevention program must be available at an organizational level to prevent politically motivated hackers, cybercrime-supported entities, and organized crime. Therefore, it is recommended to carry out the baseline assessment of the cyber-risk-level facility of operational assets.

The primary elements to be considered for such assessment are the estimation of potential damage to asset maintenance systems in the organization, the capability of technology providers to prevent and protect in case of cyberattack, organizational policies to incorporate the trials and regular audits of predictive maintenance systems and their software applications, and the availability of standard operating procedure for quick and complete isolation of the asset maintenance systems attacked within the asset maintenance network. The base level of cyber risk level is to be assessed based on these elements, and accordingly, necessary checks and balances are to be incorporated in the IPM implementation cycle phases.

Key Takeaways

- Industry 4.0 is the era of technological advancements of Internet connectivity and supported with smart systems such as the Internet of Things, i.e., access to real-time data through Cyber-Physical Systems.
- The IoT application to optimize the production and overall supply chain in the industry is referred to as the Industrial Internet of Things (IIoT).
- Predictive maintenance applications can be developed and supported through the IT infrastructure in the cloud and the digital platform, which manages the connection of assets/equipment, datasets management, and user management.
- The evolving technologies such as Robotic Process Automation, Artificial Intelligence, and the Industrial Internet of Things are the building blocks of intelligent maintenance systems, and many organizations are still exploring the utilization of these systems in maintenance processes.
- Intelligent automation systems bring together modern technologies such as cloud computing, smart analytics, mobile computing, and smart field-level automation in asset maintenance and integrate these components with the skilled workforce to produce value across the business.
- Intelligent automation systems' major components are data scientists, cognitive automation, intelligent worker, and intelligent communication tools.
- Implementing intelligent automation systems must be a strategic decision in the organization with a clear digital vision considering the long-term/short-term business objectives, the business results, and skill and process requirements for successful implementation.
- Intelligent Predictive Maintenance's key elements are the Internet of Services, Internet of Things, Smart Solutions, Smart Factory, and cloud computing.
- The key enablers of the Internet of Services are Web Service Interface and Service-Oriented Architecture. These services intend to connect people, processes, objects, and systems.
- Service-Oriented Architecture (SOA) is not a protocol, and it is a digital strategy for connecting the resource across the organization. Web services are a standard way to execute a SOA strategy.
- With the introduction of Industry 4.0 for operations-intensive industries, the organization can leverage the technologies such as Smart Solutions and Smart Factory.
- The devices/sensors of the physical assets/equipment and Information Technology systems communicate and exchange the data through the Internet using the Internet of Things technology. A conventional CMMS software integrated with the IoT automatically communicates among the connected devices/sensors/control systems without human intervention.
- Internet of Things enables workflow automation, alarms on a real-time basis, data interpretations, asset control processes, Intelligent Predictive Maintenance, cross-functional analysis, and real-time visibility.
- The system in which the various types of intelligent devices are interfaced and collaboratively communicated through smart system with an environment of Smart Factory is known as Factories of the Future.

- Machine-to-machine interface and human-machine interface in collaboration with the diagnostic and cognitive technologies are the key elements to facilitate quick decision-making and correct maintenance intervention in asset maintenance management.
- Machine learning algorithms are incorporated in the system to identify the correlations in datasets and discover unusual data patterns. The established data patterns are shown in asset maintenance predictive models.
- The IPM framework in Industry 4.0 supports businesses to achieve operations and maintenance excellence by upgrading conventional systems to intelligent embedded systems with Smart Solutions such as Cyber-Physical Systems, Internet of Things, and Internet of Services.
- The closed-loop Develop, Deploy, Integrate, Execute, Document (D^2IED) model functional configuration confirms the availability of asset/equipment historical data and predictive maintenance information of all the assets/equipment across all the organization's facilities/plants through distributed cloud network universally. It can also be interfaced with the organization's intranet.
- Even though the benefits of Intelligent Predictive Maintenance implementation are huge for asset maintenance managers, the caution of remote access of IPM systems through networks is highly vulnerable to cyberattacks, which can sabotage the business's assets. The well-organized cyberattack prevention program must be available at an organizational level to prevent politically motivated hackers, cybercrime-supported entities and organized crime.

References

Bai, H., Ma, Z., & Zhu, Y. (2012). *The application of cloud computing in smart grid status monitoring in internet of things* (Vol. 3, pp. 460–465). Springer.

Colombo, A., & Karnouskos, S. (2009). Towards the factory of the future: A service-oriented cross-layer infrastructure. In *ICT shaping the world: A scientific view* (pp. 65–81). European Telecommunications Standards Institute.

Indsutrie4.0_Report. (2013). *Recommendations for implementing the strategic initiative Industrie 4.0.* Industrie 4.0 Working Group.

Industrie4.0_Final_Report. (2013). *Recommendations for implementing the strategic initiative INDUSTRIE 4.0.*

Koch, V., Geissbauer, R., & Schrauf. (2014). *Industry 4.0 Opportunities and challenges of the industrial internet report.*

Lee, J., Kao, H., & Yang, S. (2014). Service innovation and smart analytics for Industry 4.0 and big data environment. *CIRP Proceedings, 2014*, 3–8.

Lee, J., & Wang, H. (2008). *New technologies for maintenance in complex system maintenance handbook.* Springer.

Lucke, D., Constantinescu, C., & Westkamper, E. (2008). *Smart factory a step towards the next generation of manufacturing systems and technologies for the new frontier.* Springer.

Mell, P., & Grance, T. (2011). *The NIST definition of cloud computing.*

Schaeffler Technologies AG & Co. (2015). *Report on Schaeffler Technologies.* Germany. www.schaeffler.de/en.

schroth, C., & Janner, T. (2007). Web 2.0 and SOA: Converging concepts enabling the internet of services. *IT Professional, 3*, 36–41.

Souza, M., Sacco, M., & Porto, A. (2015). Virtual manufacturing as a way for the factory of the future. *Journal of Intelligent Manufacturing, 17*(6), 725–735.

Velmurugan, R. S. (2019, May). Intelligent predictive maintenance in the context of maintenance 4.0 for oil & gas industry. *Pipeline Technology Journal, 3*, 16–22.

Yan, L., Zhang, Y., Yang, L., & Ning, H. (2008). *The internet of things: from RFID to the next generation pervasive networked systems*. CRC Press.

Yin, R. K. (2003). *Case study research design and methods*. SAGE Publications.

Zorzi, M., Gluhak, A., Lange, S., & Bassi, A. (2010). From today's intranet of things to a future internet of things: a wireless and mobility-related view. *IEEE Wireless Communications, 17*, 43–51.

Chapter 6
Asset Maintenance Strategic Plan for Operational Excellence

Abstract This chapter describes various stages of the asset maintenance strategic plan to develop and implement an asset maintenance strategy to achieve maintenance excellence. Baseline survey methodology is an essential tool that assesses the existing asset maintenance strategy and practices implementation performance through the Baseline Survey Questionnaire (BSQ). It selects a best-fit plan for a specific asset/equipment using the Analytic Hierarchy Process (AHP) technique and provides practical solutions to asset maintenance managers in operations-intensive organizations. The comprehensive AM strategic plan helps the practicing maintenance managers to assess the organization's present situation and improve upon asset maintenance strategy processes based on the guidelines obtained from the frameworks discussed in Chaps. 4 and 5. Further, implementing operational excellence as a Six Sigma project using the DMAIC approach is discussed in detail in this chapter.

6.1 Asset Maintenance Strategic Plan Development and Implementation

A practical and achievable asset maintenance strategic plan development is a critical differentiator in operations-intensive organizations. The elements of a well-prepared asset maintenance strategic plan must be based on the frameworks discussed in previous chapters, such as asset life cycle management (ALCM), asset maintenance contributing factors, and their relations with the asset maintenance framework in Chap. 4 for asset maintenance strategy and practices planning and development. Understanding the concerns among these frameworks is essential in developing an asset maintenance strategic plan. These frameworks are the basis for any asset maintenance implementation plan also. The relations among these frameworks are shown in Fig. 6.1. It shows that there is a relation in all the phases of the frameworks mentioned above. Overall, these frameworks emphasize proper planning, effective implementation, performance assessment, and the strategic plan's revisit.

R. S. Velmurugan, T. Dhingra, *Asset Maintenance Management in Industry*, International Series in Operations Research & Management Science 310, https://doi.org/10.1007/978-3-030-74154-9_6

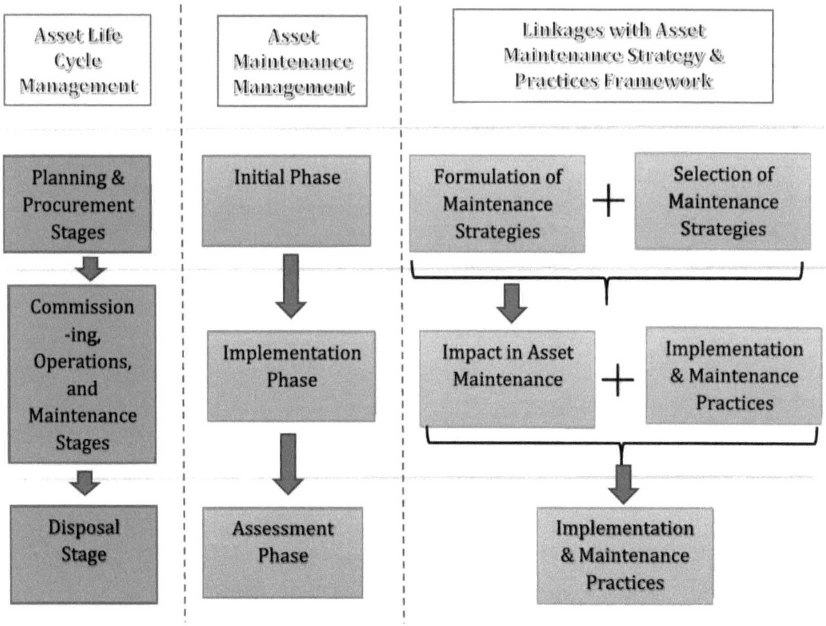

Fig. 6.1 Relationship among ALCM, AM factors, and maintenance strategy and practices framework

This chapter provides a systematic method to assess the existing asset maintenance systems available at an organization, the implementation status of asset maintenance strategies planned, the formulation of asset maintenance strategic plan and selection of best-fit plan for an organization and implementation plan and adopting the best practices of the chosen asset maintenance plan, and finally the performance evaluation of the strategic plan. It is suggested that the asset maintenance managers must follow the four-stage process, as shown in Fig. 6.2, for developing an effective asset maintenance strategic plan and its efficient implementation for achieving maintenance excellence.

The four-stage process uses the outcomes of the research study conducted among operations-intensive organizations such as asset maintenance contributing factors (Chap. 2) and asset maintenance strategy and practices planning and development (Chap. 4) for asset maintenance strategic plan implementation process. This process is explained in the subsequent sections with a method, tools, techniques, and evaluation of each stage. This process is not ending with stage 4. After implementing the strategic asset maintenance plan, the process flow goes back to stage 1 to reassess the implemented maintenance plan. The duration of the reassessment suggested in this process is on an annual basis. However, the organizations can choose such duration established on the practice by asset maintenance practitioners.

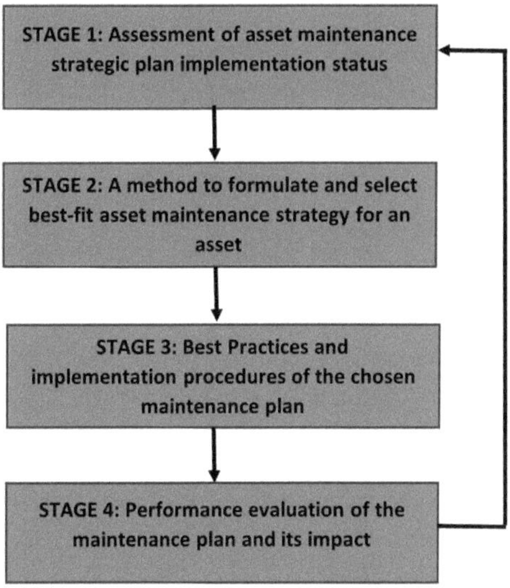

Fig. 6.2 A four-stage asset maintenance plan development and implementation

6.1.1 Stage 1: Assessment of Asset Maintenance Strategic Plan Implementation Status

The assessment of asset maintenance strategic plan execution is crucial before formulating any asset maintenance strategic plan at an organizational level. This evaluation helps asset maintenance managers/engineers understand the organization's capabilities in implementing such a scenario, challenges, and calibration required in the existing plan based on all stakeholders' feedback and essential improvements to be incorporated into the new plan. This assessment identifies the baseline of asset maintenance practices in the organization and facilitates the asset maintenance team to develop a more accurate strategic asset maintenance plan for the coming year. The suggested tool to identify the baseline of asset maintenance at an organizational level is **Baseline Survey Questionnaire**.

This section deliberates the application of the Baseline Survey Questionnaire (BSQ). It explains the method of creating such a questionnaire for conducting a semi-structured interview at an organizational level to comprehend the **stage 1** question, i.e., *what is the status of the existing asset maintenance strategic plan's execution and maintenance practices*?

The methodology for developing the Baseline Survey Questionnaire (BSQ) is based on the tool case study protocol (CSP) suggested by the *Case Study Research* book (Yin, 2003). The CSP improves the reliability of the BSQ and the study conducted in any organization. A BSQ developed for conducting data collection and analysis aids asset managers to execute the baseline survey process effectively at an organizational level. Such a questionnaire is used for single-point data collection from an individual respondent in the organization. The BSQ is a complete protocol that includes general procedures, rules, and questions based on research questions. Therefore, BSQ is not merely a survey questionnaire. The Baseline Survey Questionnaire comprises the following main sections:

- *Introduction to BSQ and its purpose*: It covers an overview of the baseline study, including the objectives, issues, and related documents about the existing asset maintenance plan in the organization.
- *BSQ interview procedure*: This section provides the objectives and qualitative questions based on the study's research questions. Therefore, the interviewer has to keep the suggested format in the BSQ while interviewing the individual respondent. Recording the potential data sources and capturing such data in proper formats are essential during the interview data collection.
- *Data collection and analysis procedure*: This procedure describes the field activities to be monitored while surveying the documentary evidence, access to the asset maintenance sites, and public sources of information. It also provides a detailed analysis procedure.
- *Outline of baseline report*: It provides the details of the baseline report such as bibliographic data, questions with data collection format, on-the-job observations, and presentation of complete documentation.

The BSQ format is based on the frameworks described in Chaps. 4 and 5 and is applied to assess the baseline of the organization's asset maintenance strategy implementation and its practices. The various sections of the Baseline Survey Questionnaire presented above are discussed in detail below from the perspective of maintenance strategy and practices of a plant/facility in operations-intensive organizations.

Introduction to Baseline Survey Questionnaire and Its Purpose (Section 1)

The background of the BSQ requires to be deliberated in this part of the questionnaire. To study the current asset maintenance strategic plan implementation and maintenance practices of a plant/facility, the critical problem statement to be kept in mind is as follows: Even though *the asset maintenance practitioners follow the structured asset maintenance plan, the cost and effect of sudden failure of an equipment/asset that happens in the plant operations led not only to production loss to the organization, but it also impacts all the stakeholders such as suppliers, customers, the safety of workers, and OEM experts. If it happens, it is required to understand the organization's capabilities in implementing asset maintenance and the challenges associated with the asset maintenance plan.* Therefore, baseline survey makes it easier to comprehend the status of asset maintenance strategy implementation and the selected organization's practices.

The central survey question for the study in BSQ is "How the asset maintenance strategic plan is developed at an organizational level for an organization's business verticals? How does the selected asset maintenance strategy for an asset/equipment be executed to maintain such asset/equipment in healthy condition and operate the plant/facility continuously with the least stoppage?"

A Baseline Survey Questionnaire is a document that contains a comprehensive procedure that provides a general rule, a data collection format for capturing inputs from the interviewee. The key objective of the BSQ in assessing asset maintenance is using a systematic approach to determine *how effectively an asset maintenance strategic plan is implemented to ensure continual operation of the plant/facility, and why a specific asset maintenance strategy is the best fit for an asset/equipment?* The additional questions that can be explored and answered through this study are as given below:

- What are the various asset maintenance strategies existing at an organization in their business areas? What are the common asset maintenance strategies among the business areas?
- What is the specific asset maintenance strategy for an asset/equipment in the business area? How the asset maintenance strategy varies among the types of assets/equipment in the plant/facility?
- What are the industry benchmarks available? To what extent the asset maintenance targets are set based on benchmark measures available in the industry?
- What is the importance of reliability-based maintenance in the organization? How the asset maintenance strategy of an organization in the industry facilitates achieving the reliability of an equipment/asset toward operations and maintenance excellence?
- What types of modern asset maintenance management methods are used in the organization for leveraging the technology?

BSQ Interview Procedure (Section 2)

This section provides guidelines for the interviewer to systematically ask the questions through a semi-structured interview while keeping track of asset maintenance contributing factors to validate all the asset maintenance management concepts. If the interviewee declares some documents, the same is to be recorded and collected from the interviewer as evidence. The complete interview is to be audio-recorded by the interviewer with permission from the interviewee.

A General Suggested Format for the BSQ Interview Guide

A: General

1. Name of the organization
2. Name of the interviewee
3. Designation
4. Location (place/country)

5. Experience (in no. of years)
6. Date of interview

B: Roles and Responsibilities of an Interviewee and the Contribution in Asset Maintenance Management Function at an Organizational Level

Interview respondent's roles and responsibilities relevant to the organization's asset maintenance function must be asked and recorded.

C-1: Asset Maintenance Strategic Plan and its Implementation at an Organizational Level

This section of the interview covers the central survey question of this baseline study, which deals with the planning and execution of maintenance strategy and practices at an organizational level. The sub-questions are formulated based on the central questions such as "How the asset maintenance strategic plan is developed for a business area and executed at an organizational level? What are the best practices adopted for implementing the asset maintenance strategic plan for the plant/facility?"

C-2: Selection of Maintenance Strategy

The additional interview questions are included in selecting a suitable maintenance strategy for an equipment/asset based on the formulated asset maintenance strategic plan. The questions developed for identifying the selected aspects of asset maintenance strategy are *Why a specific asset maintenance strategy is the best fit for an equipment/asset irrespective of the organization's business areas? What are the common asset maintenance strategies used among the business areas? How to leverage the resources by adopting common asset maintenance strategies among the organization's equipment/assets/processes? How are the asset maintenance practices managed efficiently concerning an asset maintenance strategy chosen?*

C-3: Benchmarking in Maintenance

This part of the questionnaire assesses the asset maintenance baseline in respect of benchmarking in the industry. The questions included in this part facilitate such kind of assessment at an organizational level. The questions are "What are asset maintenance targets, and how are these targets associated with the industry benchmarks? If the targets are based on the industry benchmarks, what is the level of accomplishment towards asset maintenance excellence?"

C-4: Reliability in Maintenance

This part of the BSQ covers additional baseline survey questions that deal with the reliability aspects of asset maintenance in the organization. The sub-questions are to be formulated based on the central question such as "What are the asset maintenance strategies used for achieving the asset/equipment reliability in the organizations? If such reliability-centered asset maintenance strategies are used in the organization, how does such practice facilitate uninterrupted operations and contribute to accomplishing the production targets?"

C-5: Modern Asset Maintenance Management Applications

This part of the BSQ covers additional baseline survey questions for assessing the modern asset maintenance management applications and their usage at an organizational level. The questions included in this part are "What are the modern asset maintenance methods used in asset maintenance management at an organizational level? How such new techniques, applications are beneficial to improve the asset maintenance processes by leveraging the technology?"

C-6: Asset Maintenance Strategies and Practices at an Organizational Level in All the Business Verticals—Additional Comments (Other than Above), if Any

The above parts C-1 to C-6 of the Baseline Survey Questionnaire are qualitative questions with an exploratory nature to study, understand, and assess the various asset maintenance strategy-related aspects from the employees of all the levels such as technician to the manager even with the top leaders in the organization. A sample BSQ format is provided in Appendix A.

Data Collection and Analysis Procedure (Section 3)

The next logical step shall be data collection and the procedure to be adopted in collecting the data through the semi-structured interview process from the maintenance managers/engineers in the selected organization. If one or more business verticals of the organization are taken for the baseline survey, then each business vertical must be considered an individual case. Data collection and analysis of each business vertical are to be conducted separately at an organizational level. A list of employees who can be interviewed must be chosen across the levels/grades of employees to cover the entire spectrum of the employee grades from technician to manager. Further, the selected employees' location and contact details must be prepared before starting the semi-structured interview process. Accordingly, site visits can be planned to include all the selected interview respondents one by one, and participant observation is also to be carried out to understand the detailed asset maintenance implementation processes. The expected prerequisites before conducting the semi-structured interview are as follows:

- The permission and date of interview from the individual respondent, i.e., an employee in the organization, need to be obtained well in advance.
- A day before an interview, intimation is to be given to the concerned interviewee regarding the interview details such as timing, duration, and overview.

- The interviewer must be ready with the BSQ protocol procedure and the essential requirements such as BSQ format, relevant documents, audio recording device, and a brief on the process.

The semi-structured interviews conducted are audio-recorded during the interview. Therefore, the necessary intimation must be given to the concerned maintenance manager/engineer in the organization regarding the interview schedule and methodology, and necessary permissions are to be obtained well before starting the interview process. A sample Baseline Survey Questionnaire is provided in Appendix A. Further, the audio recordings of semi-structured interviews are transcribed, and the data analysis can be done through any suitable qualitative software (e.g., Atlas. ti). A detailed data analysis procedure is provided in Appendix B.

The above study's key objective using the BSQ approach is "being best in the industry is not about benchmarking of maintenance practices within the industry or benchmarking with other industries but a clear understanding about our asset maintenance strategy and practices and learnings out of such kind of study." One case study using this approach is provided in Appendix C.

Outline of the Baseline Report (Section 4)

The baseline report presents the data analysis findings with supported qualitative analysis software outcomes such as keywords, key phrases, and network diagrams. The findings are emerging out from the interview respondents within the organizations only. Therefore, such inputs from the employee are valuable in understanding the efficiency of asset maintenance strategic plan implementation and suggestions for improvement in asset maintenance plan so that efficiency can be leveraged toward asset maintenance excellence.

6.1.2 Stage 2: A Method to Formulate and Select Best-Fit Asset Maintenance Strategy for an Asset

Analytic Hierarchy Process (AHP) and sensitivity analysis mathematical tools can be used to decide on asset maintenance strategy selection. AHP technique can be applied to solve complex problems in the domain of asset maintenance management. These problems can have different objectives, and they may have correlations among these objectives. Therefore, AHP helps to solve complex decision-making problems by defining hierarchy levels for the chosen objectives with the appropriate categorization of criteria and sub-criteria, leading to various options for the problem (Saaty, 2008).

The measurement of AHP is based on pairwise comparisons and depends on subject matter experts (SME). The priorities are identified among all the sub-criteria and criteria using the AHP technique in all the levels. Therefore, the industry practitioners use this method for various industrial applications, such as selecting an asset maintenance strategy since AHP is reliable. In this section, a methodology developed to choose a maintenance strategy among the available strategies is presented in detail.

The AHP modeling process includes four stages such as defining the problem, constructing the hierarchy of the formulated decisions, developing comparisons (pairwise) for each decision matrix, and using the priorities by weightage from the available comparisons for a level, and several iterations are carried out to choose among the alternative by the allocating the weightage priorities till achieving the absolute priorities (Saaty, 2008). The steps involved in the AHP iteration process are as given below:

1. Problem definition and its objectives
2. Identification of criteria and sub-criteria for the objective chosen
3. Construction of the alternatives for the problem defined
4. Development of a hierarchy framework for analysis
5. Collection of empirical information and data
6. Preformation of pairwise comparisons for each level of criteria and sub-criteria
7. Conduction of the consistency test
8. Calculation of the comprehensive weight parameters of each criterion and sub-criteria
9. Synthesizing the results
10. Sensitivity analysis
11. The concluding ranking of proposed alternatives

This section aims to provide a methodology to maintenance practitioners regarding formulation, selection, and implementation of asset maintenance strategy at an organizational level using the Analytic Hierarchy Process. It is explained with two problem definitions regarding asset maintenance strategy.

Problem I: To evaluate and select the most important elements contributing to asset maintenance strategy formulation

Problem II: To assess and choose the best asset maintenance strategy for the process/asset/facility

Referring to the steps involved in the AHP process as mentioned above, the first 1–4 steps, i.e., up to the development of framework developed for the above problems, are shown in Figs. 6.3 and 6.4 and explained in this chapter as a guide to maintenance managers/engineers. Therefore, maintenance managers/engineers can utilize the given hierarchy frameworks as an example, and they can conduct this study at their organizations to arrive at the best maintenance strategy selection for the asset/process/equipment. Further, an explanation of all the remaining steps is given below.

The next step, i.e., step 5, is nominating a team of assessors from the asset maintenance team for data collection and measurement after developing the AHP hierarchy framework. The assessors can also be selected among all the stakeholders

Level 1: Objective

Maintenance Strategy Formulation Evaluation

Level 2: Criteria

1.Maintenance Tactics; 2. Reliability Analysis; 3. Maintenance Performance; 4. Materials Management; 5. Planning & Scheduling; 6. Information Technology; 7. Organization/ HR Resources; 8. Maintenance Policy /Budget; 9. Maintenance Challenges; 10. Employee Empowerment

Level 3: Sub-criteria

1. a) Preventive Maintenance; b) Predictive Maintenance; c) Total Maintenance Hours; d) Compliance of Maintenance Schedules; e) OEM Recommendations; f) Reliability Centred Maintenance; g) Shutdown maintenance; h) Total Productive Maintenance
2. a) Equipment History Documentation; b) Critical Equipment Identification; c) Root Cause Analysis; d) Meantime Between Failures; e) Maintenance Value Risk Study; f) Asset Reliability Statistics
3. a) Maintenance Cost; b) Equipment History; c) Training Manhours; d) Internal/ Industry Norms; e) Benchmarking/Targets
4. a) Stores Management; b) Purchase Process; c) Stock management & materials scheduling; d) Optimization & Rationalization of Material Process
5. a) Equipment History; b) Work order with priority; c) Preventive Maintenance Schedules; d) Weekly work schedules; e) Long-term/Shutdown schedules; f) Measurement work backlog
6. a) Computerized Maintenance Management System; b) Integrations of Maintenance & Materials Management; c) Condition Based Maintenance with automation; d) Expert Systems for Diagnosis
7. a) Maintenance Staff Level; b) Maintenance Organization; c) Responsibility of Firstline Supervisors; d) Technical Training; e) Apprenticeship Program; f) Deployment of Contractors
8. a) Maintenance Mission & Objectives; b) Integration of Maintenance plan & Business Plan; c) Maintenance Policies; d) Maintenance Budget
9. a) Technical Issues; b) Contract Issues; c) Stake holder issues; d) Internal Issues; e) Unexpected issues
10. a) Multi skilled people; b) Operator Based Maintenance; c) Regular interaction with maintenance staff; d) Self-directed work teams; e) Change management procedures; f) Partnership with key suppliers/contractors

Level 4: Alternatives

1. Preference over Preventive Maintenance Vs Predictive Maintenance; 2. Shutdown maintenance is critical over other maintenance tactics; 3. RCM and Reliability Analysis preferred over others maintenance tactics; 4. Human Resources & Employee empowerment are key elements in strategy formulation; 5. Maintenance Strategy (MS) formulation should incorporate envisaging maintenance challenges; 6. MS formulation should focus on materials management/planning & scheduling; 7. Implementation of information technology impacts maintenance strategy formulation

Fig. 6.3 Hierarchy framework for maintenance strategy formulation evaluation

Level 1: Objective

Maintenance Strategy Selection

Level 2: Criteria

1. Maintenance personnel experience; 2. Asset criticality & availability; 3. Process complexity & severity; 4. OEM recommendations; 5. Maintenance Policy guidelines

Level 3: Sub-criteria

1. a) Culture of the organization, b) Size of the organization, c) Maintenance objectives, d) Behaviour of the specific process and equipment in the country/region
2. a) History records of preventive maintenance, b) Breakdown maintenance of the specific asset, c) Mean Time Between Failures, d) failure probability index, e) spare procurement lead time, f) Asset replacement value, g) OEM recommendations, safety and environment impact, h) Ability to isolate single-point failures, and stake holders impact
3. a) Mitigation of plant risks, b) Health & safety, environmental impact, c) Local government laws and rules, d) Emergency & disasters, and e) Technological upgradation of the process.
4. a) Preventive maintenance after specific running hours of equipment, b) Replacement of materials/ spares periodically, c) Standard maintenance practices
5. a) Capacity utilization, b) Maintenance Cost, c) Production cost, d) Spares consumption, e) Product quality, f) Internal norms, g) Industry norms

Level 4: Alternatives

1. Specific maintenance strategy for a particular asset/process; 2. Application of Maintenance Decision Support System; 3. Maintenance strategy based on OEM recommendations; 4. Modern maintenance strategies such as Reliability Centered Maintenance and Risk Based Maintenance; 5. Intelligent Predictive Maintenance over traditional maintenance tactics

Fig. 6.4 Hierarchy framework for maintenance strategy selection

such as operation managers/engineers, maintenance managers/engineers, and other support service managers/engineers at an organizational level for a business vertical chosen for the organization.

The next step, i.e., step 6, is a pairwise comparison of criterion and sub-criterion chosen for the problem's defined objective. The assessor's team must be trained on a pairwise comparison method before their assignment in the assessment process. The objectives shown in Figs. 6.3 and 6.4 are taken as the basis for pairwise comparison by assigning the relative scale values to the criteria/sub-criteria at each hierarchy level.

A geometric mean approach is used at each hierarchy level to combine the decisions to obtain the relevant consensus. An intensity of importance scale (1–7) is used for ranking (refer to Table 6.1). The ranking process helps to identify the

Table 6.1 Scale of relative preference for pairwise comparison

Level of influence (scale)	Decision	Description (in comparison with one factor with another)
1	Equal ranking	Equal influence on the objective
3	The ranking is relatively higher	Influence is higher than the other based on the experience
5	The ranking is relatively much higher	Influence is much higher than the other based on the experience
7	Ranking is significant	There is a significant influence based on the experience and strong demonstration that exists
9	Ranking is critical	The factor influences the process critically with the highest possible contribution
2, 4, 6, 8	Intermediate values	When a concession is required between the two contiguous decisions

no. of times one criterion is chosen over other criteria. Similarly, one sub-criteria is selected over another sub-criteria. Accordingly, the most preferred decision can be chosen from the alternatives in respect of the problem statement objective (Saaty, 2008).

Once the pairwise comparison table is prepared, it is required to check whether the generated pairs of the criteria/sub-criteria are coherent or not. The same is calculated using the formula Consistency Ratio (CR). The Consistency Ratio should be below 10%. In case CR is greater than 10%, then possible reasons are to be reexamined. Further, to ensure consistency, a simple Shewhart X-bar-R chart or exponentially weighted moving average (EWMA) chart is used. Therefore, the Consistency Index (CI) can be found from the following equation:

$$\mathrm{CI} = (\lambda_{\max} - n)/(n - 1)$$

where n = no. of criteria/sub-criteria at each level

$\lambda_{\max}$ = highest matrix eigenvector

A suitable statistical software package can be used to calculate CR and CI for the taken problem. The conduction of the consistency test is the seventh step. In the eighth step, the global weights for each criterion and sub-criterion are calculated with local weights assigned according to the previous hierarchical level. The global weights are considered the highest hierarchical level in the process that provides ranking among the elements of each sub-criteria/criteria and the ranking of all sub-criterion/criterion elements. The last step is obtaining the results by using the single-decision criteria method, i.e., all the alternatives are multiplied by the global weight.

6.1.3 Stage 3: Best Practices and Implementation Procedures of the Chosen Asset Maintenance Plan

Operations-intensive organizations are making many efforts to standardize their asset maintenance implementation. These organizations' key elements are industry standards, internal organizational policies, and peer best practices. Industry standards available in the domain of asset maintenance management are discussed below. A nine-step process for achieving the best asset maintenance practices is discussed in detail in Chap. 3. A chronological list of standards used in asset maintenance management and asset performance evaluation is shown in Fig. 6.5.

ISO 9000 is a standard for quality management developed in 1987, and many organizations follow this standard in the industry until now. This standard aims for significant objectives such as customer satisfaction and business efficiency. ISO 9000 consists of several clauses focusing on maximizing production, reducing unwarranted expenditures, and ensuring quality in processes and products. ISO 9000 helps organizations improve the documentation and retrieval of documents with traceability, but a gap exists in assessing the asset maintenance performance and managing the assets effectively throughout the asset life cycle.

Industry standards (e.g., Publicly Available Specification (PAS) 55 and ISO 55000) are an aid to achieve the best asset maintenance practices systematically. A PAS 55:2004 asset maintenance management standard was developed by the Institute of Asset Management (IAM) and the British Standards Institution (BSI). Further, 50 multinational organizations from various 15 industry sectors in ten countries established a Publicly Available Specification (PAS) 55:2008 updated standard for physical asset maintenance management. This standard has been accepted and implemented in most industry sectors in public/private sector organizations and is also part of practice in various environments, regulatory systems, and cultures. The primary benefit of applying this standard is that it provides a well-defined process for organizing all asset maintenance endeavors in general toward business excellence, i.e., bottom-line performance improvement in the organization's financials. The International Standards Organization (ISO) then recognized PAS 55 as the source for advancing the ISO 55000 series of international standards.

BS ISO 55000:2014 standard is available in three parts. ISO 55000: Asset Management—Overview, Principles, and Terminology helps develop long-term asset maintenance plan. ISO 55001—Asset Management Systems and Requirements is for improvement in asset maintenance management. ISO 55002 offers recommendations for the use of ISO 55001. In all these standards, the conditions are stated generically to facilitate various organizations' types and sizes to adopt the

Fig. 6.5 Standards in chronological order in respect of asset maintenance management

procedures. Therefore, asset maintenance managers/engineers need to thoroughly study the standards and interpret the standard's requirements to fit an organization and its operating environment. If a specific maintenance strategy is chosen from the above stages 1 and 2, standards can be used to systematically implement such a process by fulfilling the standards' requirements.

The other standards, such as ISO 50000:2011 on energy management, ISO 27001:2013 on information security, and ISO 45001:2016 on occupational health and safety, facilitate the asset maintenance processes implementation by assuring achieving the organizational business objectives consistently and sustainably over the period. Therefore, standards are performing a decisive role in asset maintenance strategic plan implementation. A systematic implementation procedure for operations-intensive organizations in the industry is as shown in Fig. 6.6.

An asset maintenance strategy is chosen from the stage 2 process for an asset/equipment/plant/facility. The asset maintenance plan must be based on the selected asset strategy and asset maintenance internal norms at an organizational level. This maintenance plan is asset-specific. For planning and scheduling purposes of the asset, maintenance plan is categorized as long-term or short-term plans. Such categorization helps to allocate budget and necessary approvals from corporate management.

Further, asset maintenance managers can work on resources, operational requirements, spare identification, and availability. Once this step is completed, job order creation and timely notification for the execution of asset maintenance plans are scheduled through CMMS/ERP/EAM systems. After completing the asset maintenance job, proper documentation is the must activity of asset maintenance engineers/technicians. Job performance evaluation and quality checks are the built-in modules in ERP/EAM systems. In case the job cannot be executed due to the nonavailability of the asset/equipment due to operational requirements. Necessary approvals are to be accorded from the concerned authority like the plant manager, and proper documentation must be maintained as per standards. The complete flow asset maintenance strategic plan needs to be supported with the standard application at an organizational level.

Job order creation and execution are crucial to the implementation of an asset maintenance plan. The systematic process of job order helps asset maintenance managers/engineers/technicians to organize, assign, prioritize, track, and complete all the tasks related to a job order. If this process is followed well, job order information available in the asset management systems (ERP and EAM) enables efficient execution of job orders. The job order process includes the following steps (as depicted in Fig. 6.7): task identification, job order creation, job order prioritization and scheduling, assignment of the job to technician/engineer and its completion, closing of the job order in the EAM/ERP system and documentation, and job order analysis for identification of any rework or approval for deviation.

The types of asset maintenance tasks are planned maintenance jobs like preventive maintenance schedules and unplanned maintenance jobs like asset/equipment breakdown jobs. Accordingly, the task identified is to be defined in the first step for further planning in the job order cycle. In the case of breakdown jobs, the operation

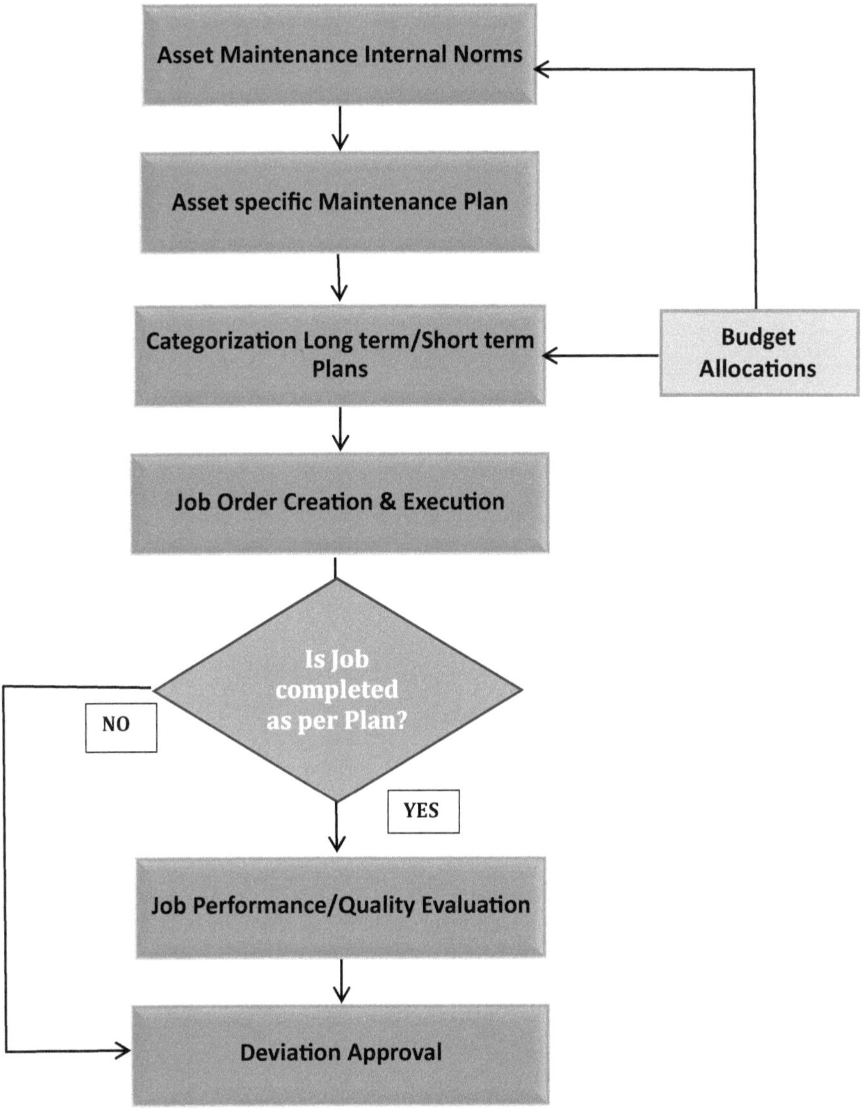

Fig. 6.6 Asset maintenance plan implementation flow chart

engineer creates the job order and forwards it to maintenance engineers through the EAM system.

Otherwise, in planned maintenance jobs, the job order is created automatically and triggered to both operations and maintenance engineers at the proper time as per schedule. The asset maintenance manager prioritizes the jobs based on operations requirements, spares, and workforce availability. Accordingly, scheduling is done based on planned maintenance triggers, timelines, and job execution permits from

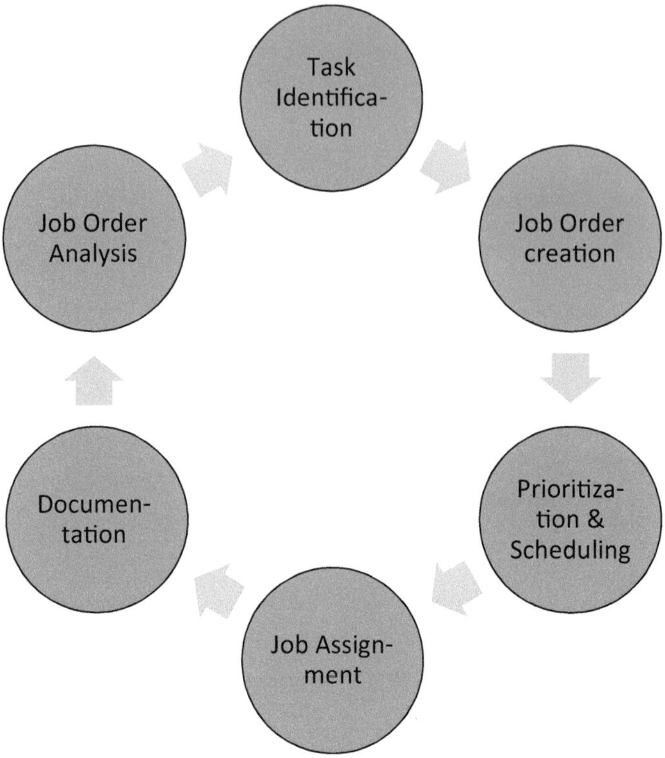

Fig. 6.7 Job order cycle for asset maintenance jobs

operators. Upon completion of the job and job order, logging by maintenance engineer/technician is a critical aspect in asset maintenance for developing asset/equipment history, reviewing the past job solutions, and preparing for technical audits. Job order documentation contains valuable information, and such documentation provides insights about asset maintenance systems and processes. The necessary information can be retrieved quickly from the system if the failure reoccurs again.

The entire asset maintenance plan workflow can be automated from field equipment to human interface using the solutions offered by vendors based on clouds such as SaaS applications with mobile connectivity so that the operator can capture the failure from the plant at the point of work. Such digitized solutions are helpful to asset maintenance managers to convert the asset maintenance transaction to management metrics.

6.1.4 Stage 4: Performance Evaluation of the Maintenance Plan and Its Impact

Performance evaluation of the asset maintenance plan is critical for asset maintenance strategic plan implementation in operations-intensive organizations. Detailed evaluation criteria and best practices in the industry are given in Table 6.2.

A good asset maintenance strategic plan must include the above performance factors and best practices built-in to the plan. The asset maintenance managers should keep in mind that such performance measurements should be included while preparing the maintenance plan.

Table 6.2 Performance evaluation criteria for the asset maintenance plan

Sl. no.	Evaluation criteria	Best practices
1	Maintaining asset/equipment inventory includes the following: unique asset tag ID, type, condition, specifications, asset service history, criticality, insurance details, remaining useful life, and replacement cost	Asset registry or utility modules are available as standard in ERP/EAM systems to maintain asset inventory
2	Identification of asset location using unique ID for remote assets to capture asset information	Asset location identification is maintained with the help of GIS systems and geolocation indicators
3	Assessment of asset condition and its performance for all the assets in a plant/facility	Condition assessment rating systems as recommended by industry standards/internal norms at an organizational level
4	Availability of a list of asset failure history and ranking of assets according to the likelihood and consequences of asset failure	Condition assessment rating systems as recommended by industry standards/internal norms at an organizational level
5	Increasing asset maintenance cost	Critical monitoring and quarterly review of asset maintenance cost as planned cost vs. actual cost and identify the root cause reasons for avoiding its recurrence
6	Maintenance contracts workforce overruns	Strict compliance of contract workforce utilization is as per the asset maintenance plan
7	Premature or unexpected asset/equipment failure	Regular compliance audits and revisiting the asset maintenance plans must be standard practice
8	Customer satisfaction	Track the asset maintenance team workflow, job completion, and resolution status and regular meetings with all stakeholders

6.2 Asset Maintenance for Operational Excellence

Operational excellence strives to achieve a reduction in resource costs and maximizes the performance of the production process. Maximizing the performance means in operations is production capacity at an optimum level and reduces the production loss and waste. Operational excellence can be achieved by ensuring equipment/asset availability, increasing asset reliability, maximizing production yield with prescribed quality, delivering products as per market requirement, and maintaining safety performance with sustainability standards. The other element that plays a crucial role in operational excellence is the reduction in resource costs. The cost reduction can be achieved by reducing energy consumption, optimizing raw material consumption and inventory levels, and reducing workforce costs and capital resources.

The suggested process improvement tool is the Lean Six Sigma DMAIC approach to achieve operational excellence. This approach consists of five phases such as define, measure, analyze, improve, and control. Operational excellence must be treated as a project. The phase-wise application of suggested steps, tools by Six Sigma, and the operational excellence project inferences from this approach are shown in Figs. 6.8, 6.9, 6.10, 6.11, and 6.12.

The operational excellence project is majorly dependent on asset maintenance excellence, as discussed earlier in earlier chapters. Once the as-is status of the operation is assessed, the project's content and scope are prepared. **Process optimization** is the critical objective in the operational excellence—define phase (Fig. 6.8) of operational excellence, and in turn, process optimization is based on process control technology. The management of process control is with the asset maintenance managers in operations-intensive organizations. Process control uses

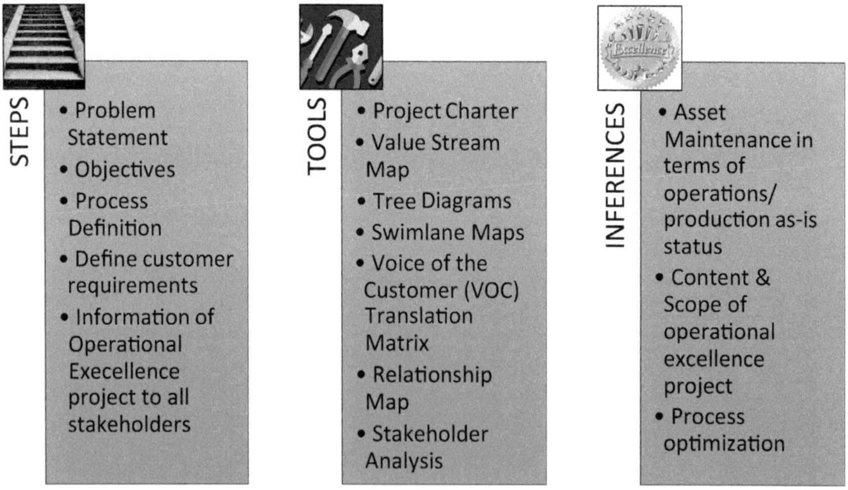

Fig. 6.8 Operational excellence—define phase

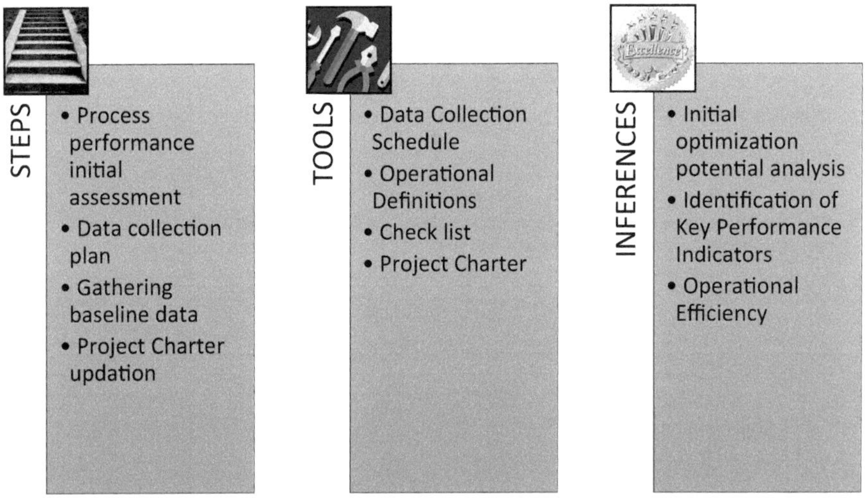

Fig. 6.9 Operational excellence—measure phase

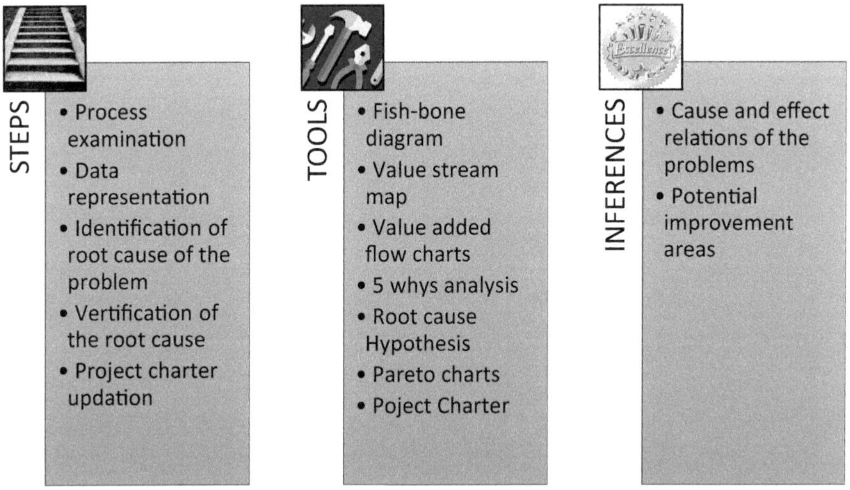

Fig. 6.10 Operational excellence—analyze phase

automation systems and field instrumentation to manage the process. Process control is combined with process models that accurately capture the production and logistic flows. The next phase in the operational excellence project is the measure phase.

In the measure phase (Fig. 6.9), identifying KPIs is important to measure the operational excellence improvement after assessing the optimization potential analysis. The operational excellence project is taken up only if there is a scope available for improvement in the process. The KPIs must be derived based on the significant

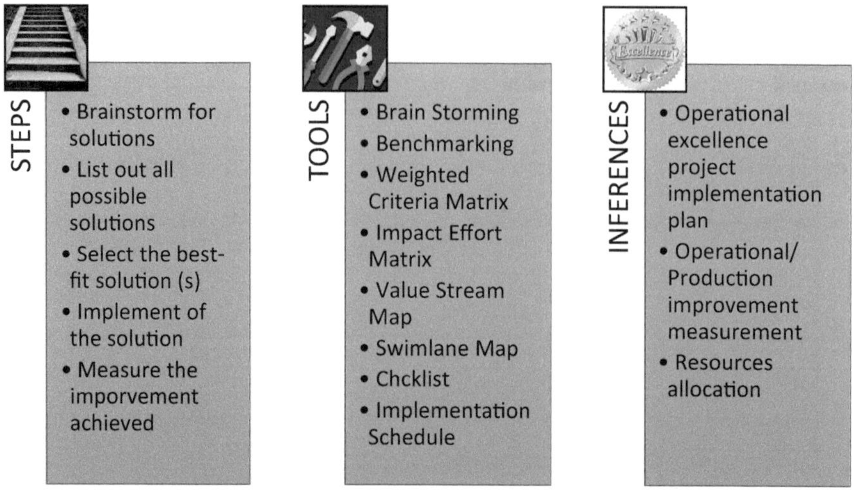

Fig. 6.11 Operational excellence—improve phase

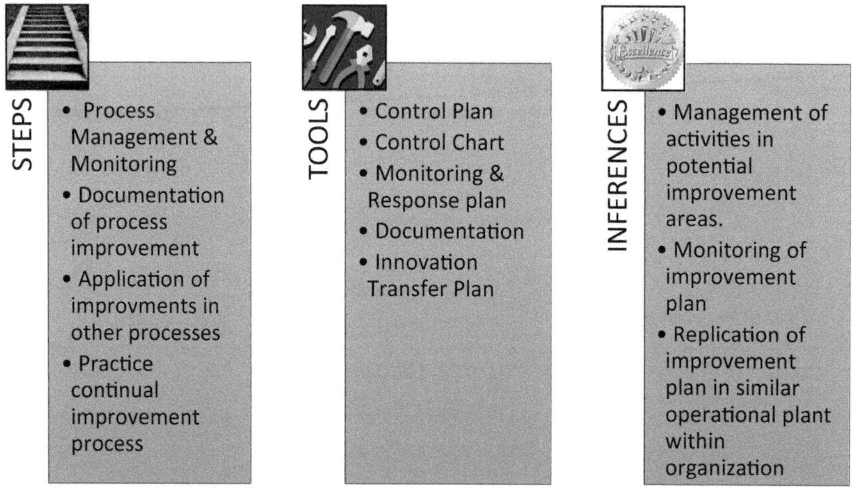

Fig. 6.12 Operational excellence—control phase

cost-consuming area and potential efficiency improvement opportunities to achieve operational efficiency.

The next logical step (Fig. 6.10) of the operational excellence project analyzes the measured data related to the operations/production. The root cause of the operational issues is to be arrived at based on the data. Once the root cause is identified, the problem's cause must be verified with the major stakeholders such as operational managers and asset maintenance managers to confirm the outcomes. The final cause and effect relations are to be prepared and the potential improvement areas for achieving operational excellence.

The fourth step in the Six Sigma DMAIC process is the improve phase as shown in Fig. 6.11. From the analyze phase's inputs, the operational excellence project implementation plan is prepared for all the potential improvement areas for achieving operational efficiency. Few such improvement areas are listed out for the benefit of asset maintenance managers and production managers. The areas are reducing energy consumption, increasing the production target, reducing inventory, reducing the lead time of products shipping or transportation, fewer off-spec products, reducing plant start-up time for production, and higher asset/equipment availability.

Thorough measurement of KPIs defined in the measure phase for all such identified potential improvement areas and their documentation is essential for managing and controlling the operational performance. Improvement measures require resources such as financial budget, material, and workforce. Therefore, resource allocations for improvement activities are crucial for achieving operational excellence.

The last phase (Fig. 6.12) of the operational excellence project is the control phase. This phase's primary aim is managing the improvement-plan-related activities implementation and continuous monitoring of all the associated activities to avoid slackness. If the improvement plan is beneficial, such a plan can be implemented in other similar process/operational areas within the organization. The overall project success depends on the quick adoption of solutions in the existing workflows, onboard operations/maintenance engineers, and their training to monitor and control the implemented functions.

Key Takeaways

- The assessment of asset maintenance strategic plan execution is crucial before formulating any asset maintenance strategic plan at an organizational level.
- The suggested tool to identify the baseline of asset maintenance at an organizational level is Baseline Survey Questionnaire (BSQ).
- A Baseline Survey Questionnaire is a document that contains the questionnaire (instrument) for data collection, the procedure, and general rules to be followed in using the BSQ.
- BSQ serves as a guideline specifying the type of questions in conducting the semi-structured interview for the baseline survey. The asset maintenance contributing factors are used for validation purposes during the interview.
- The study's key objective using the BSQ approach is "being best in the industry is not about benchmarking of maintenance practices within the industry or benchmarking with other industries but a clear understanding about our maintenance strategy and practices and learnings out of such kind of study."
- Analytic Hierarchy Process (AHP) and sensitivity analysis mathematical tools can be used to decide on the maintenance strategy selection.
- AHP helps solve complex asset maintenance decision-making problems by defining different hierarchy levels for the defined objective/goal with suitably categorized criteria and sub-criteria leading to various alternatives.

- ISO 9000 helps organizations improve the documentation and retrieval of documents with traceability, but a gap exists in assessing the asset maintenance performance and managing the assets effectively throughout the asset life cycle.
- Industry standards such as PAS 55 and ISO 55000 help achieve the best asset maintenance practices systematically.
- Job order creation and execution are crucial to the implementation of an asset maintenance plan. The systematic process of job order helps asset maintenance managers/engineers/technicians to organize, assign, prioritize, track, and complete all the tasks related to a job order.
- The job order process includes the following steps: task identification, job order creation, job order prioritization and scheduling, assignment of the job to technician/engineer and its completion, closing of the job order in the EAM/ERP system and documentation, and job order analysis for identification of any rework or approval for deviation.

References

Saaty, T. L. (2008). Relative measurement and its generalization in decision making why pairwise comparisons are central in mathematics for the measurement of intangible factors the analytic hierarchy/network process. *RACSAM-Revista de la Real Academia de Ciencias Exactas, Fisicas y Naturales. Serie A. Matematicas, 102*(2), 251–318.

Yin, R. K. (2003). *Case study research design and methods*. SAGE Publications.

Appendix A

Baseline Survey Questionnaire (BSQ)

A. General

1. Name of the Company:
2. Name:
3. Designation:
4. Location:
5. Experience:
6. Date of Interview:

B. Roles and Responsibilities Relevant to Asset Maintenance Function in the Organization

____

____

____

____

C-1. Asset Maintenance Strategic Plan and Its Implementation at an Organizational Level

This section of the interview covers the central survey question of this baseline study, which deals with the planning and execution of maintenance strategy and

R. S. Velmurugan, T. Dhingra, *Asset Maintenance Management in Industry*,
International Series in Operations Research & Management Science 310,
https://doi.org/10.1007/978-3-030-74154-9

practices at an organizational level. The sub-questions are formulated based on the central questions such as *"How the asset maintenance strategic plan is developed for a business area and executed at an organization level?" "What are the best practices adopted for implementing the asset maintenance strategic plan for the plant/facility?"*

How the organization's policies/objectives influence the asset maintenance function in the organization?

____

____

Describe the maintenance organization/structure and its roles in maintenance in your organization?

____

____

Describe your asset maintenance mission and objectives?

____

____

Do you have a maintenance policy for your organization? If yes, please describe.

____

____

What are the factors that impact maintenance strategy planning? How do you manage to prepare asset maintenance strategies for different business portfolios?

____

____

What management approaches are adopted in developing the organization's asset maintenance strategic plan for different business portfolios?

____

____

How have the maintenance strategies influenced the maintenance budget in your organization? How do you formulate the maintenance budget?

____

____

How do you manage to implement the maintenance strategies in your organization?

____

____

What are the driving forces and challenges in asset maintenance practices?

____

____

How do you monitor the execution of maintenance strategies in your organization?

____

____

Do you envisage employee empowerment becoming important for maintenance function?

____

____

Are you using operator-based maintenance/autonomous maintenance? If yes, is such type of maintenance practiced in your organization?

____

____

How do you interact with your work teams/supervisors in the maintenance function?

____

____

____

How do you manage modification works carried out by the maintenance teams in your organization?

____

____

How is the maintenance function influenced by external stakeholders such as key suppliers/contractors?

____

____

How do you envisage materials management influence in maintenance management?

____

____

Do you foresee that training impact the execution of maintenance strategies and maintenance practices?

____

____

How has information technology been leveraged in the formulation and implementation of maintenance strategies?

____

____

How frequently your plant faces shutdown? How do you ensure the smooth operation of the plant?

____

____

How has planning and scheduling influenced the maintenance practices in your plant?

____

____

How do you manage long-term plans to forecast major maintenance works/ shutdowns in your plant?

____

____

How do you manage short-term plans to comply with the plant's preventive maintenance works/stoppages?

____

____

How do you prioritize the day-to-day maintenance works/backlog maintenance works?

____

____

What are the management approaches to leverage Information Technology (IT) in maintenance function in your plant?

____

____

Do you foresee any material required for the execution of a particular maintenance strategy? If yes, how do you manage the inventory of material in your plant?

____

____

How has inventory control influenced maintenance planning?

____

____

C-2. Selection of Maintenance Strategy

The additional interview questions are included in selecting a suitable maintenance strategy for an equipment/asset based on the formulated asset maintenance strategic plan. The questions developed for identifying the selected aspects of asset maintenance strategy are as follows: *"Why a specific asset maintenance strategy is the best fit for an equipment/asset irrespective of the business areas in the organization?" What are the common asset maintenance strategies used among the business areas? How to leverage the resources by adopting common asset maintenance strategies among the organization's equipment/assets/processes? How are the asset maintenance practices managed efficiently concerning an asset maintenance strategy chosen?*

Describe the asset maintenance strategies and practices in your plant. How do you come up with such maintenance strategies?

____

____

Describe the maintenance strategies adopted for a particular process/equipment within your plant. Are they the same or different from the type of process/equipment?

____

____

What decision processes is your organization applying to select particular maintenance strategies for a particular process/equipment? How do they impact the overall maintenance practices in your plant maintenance?

____

____

How do you manage total maintenance man-hours to the particular maintenance strategy (e.g., preventive maintenance)?

____

____

How do you manage compliance with maintenance strategy schedules (e.g., preventive maintenance)?

____

____

How has the emergency/breakdown maintenance influenced the devotion of total man-hours in your plant?

____

____

Compare and describe the maintenance strategies and maintenance practices adopted in various plant operations.

____

____

How would you characterize maintenance strategies and practices among the different plant operations?

____

____

Describe the maintenance strategies and practices adopted for the various equipment/processes available within the organization.

____

____

How would you characterize maintenance strategies and practices among other business verticals of your organization?

____

____

C-3. Benchmarking in Maintenance

This part of the questionnaire assesses the asset maintenance baseline in respect of benchmarking in the industry. The questions included in this part facilitate such kind of assessment at an organizational level. The questions are as follows: *"What are asset maintenance targets, and how are these targets associated with the industry benchmarks? If the targets are based on the industry benchmarks, what is the level of accomplishment toward asset maintenance excellence?"*

How is the maintenance performance of "best in class" benchmarked at your organization?

____

____

Describe the benchmark measures and targets in practices at your plant.

____

____

What initiatives and processes are being used to achieve the benchmark parameters of your plant in practices?

____

____

How do you manage the organization's internal norms and industry norms in maintenance performance measurement/benchmarking?

____

____

C-4. Reliability in Maintenance

This part of the BSQ covers additional baseline survey questions that deal with the reliability aspects of asset maintenance in the organization. The sub-questions are to be formulated based on the central question such as *"What are the asset maintenance strategies used for achieving the asset/equipment reliability in the organizations? If such reliability-centered asset maintenance strategies are used in the organization, how much practice facilitates uninterrupted operations and contributes to accomplishing the production targets?"*

How do plant operations cope with the reliability issues, and what is the maintenance strategy adopted to ensure plant operations' reliability specific to particular processes/equipment/assets available in your plant?

____

____

How are the reliability statistics of process/equipment/asset maintained in your plant?

____

____

What initiatives and processes are used to do incident/root cause analysis of plant shutdowns and equipment failures?

____

____

How do you manage with reliability-based analysis of process/equipment/asset being maintained in your plant?

____

____

What is the major challenge in maintenance strategy formulation and maintenance practices in your organization?

____

____

C-5. Modern Asset Maintenance Management Applications

This part of the BSQ covers additional baseline survey questions for assessing the modern asset maintenance method applications and their usage at an organizational level. The questions included in this part are as follows: *"What are the modern asset maintenance methods used in asset maintenance management at an organizational level? How are such new techniques and applications beneficial in improving asset maintenance processes by leveraging the technology?"*

Describe the modern asset maintenance approaches used in your organization.

____

____

How do you align Industry 4.0 guidelines in Maintenance 4.0 practices in asset maintenance management?

____

____

Do you believe that the digital transformation has helped improve asset maintenance activities toward achieving maintenance excellence?

____

____

How do you implement the recent Operations Research and Management Science developments in your organization's asset maintenance function?

____

____

C-6. Asset Maintenance Strategies and Practices at an Organizational Level in All the Business Verticals

Anything to share more (closing)

The above questionnaire is a sample, and the questions are developed to explore all the aspects of asset maintenance function in the industry. The questions aid the interviewer in conducting semi-structured interviews among the participants chosen for the interview. This questionnaire is not for distribution, and individual participant (interviewee) is not required to fill. The interview using this questionnaire may be lost for 45–60 min. The interview requires to be audio-recorded and transcribed. The transcribed text file is used for data analysis. The details of data analysis using an example are given in Appendix B.

Appendix B

Data Analysis Procedure

The transcription of the semi-structured interview conducted for asset maintenance strategy and practices using the Baseline Survey Questionnaire is based on data analysis. Data analysis must aim to understand the organization's asset maintenance strategy and practices, including benchmarking and reliability maintenance practices, asset maintenance selection methods and processes, the relationship between critical contributing factors for asset maintenance, and the asset maintenance processes existing in the organization. Every plant operations or similar plant operations are to be considered as a single case study. If different plant operations or business vertical needs to be studied, it must be treated as a separate case. Two business verticals of the single organization are considered for the baseline survey. The cross-case analysis between the verticals can understand similarities and differences among the asset maintenance processes.

After completing data using BSQ, text analysis must be carried out to identify the categories, codes, themes, and processes manually or by software like Atlas.ti. Further, patterns are also required to be identified to describe the processes captured from the asset maintenance practitioners for understanding the existing asset maintenance planning and implementation level of the organization. Data analysis can be carried out with textual analysis and cross-case synthesis. Textual analysis can be done with Atlas.ti software, network diagrams with the necessary quotation, and frequency analysis.

Interview script coding can be done through the qualitative analysis software Atlas.ti. This software enabled the data analysis process by coding, linking codes and text segments, creating memos, searching, editing, reorganizing, and visually displaying data and findings. An example is shown in Fig. B.1, which depicts the interlinking of concepts, categories, and codes for understanding the organization's asset maintenance processes.

R. S. Velmurugan, T. Dhingra, *Asset Maintenance Management in Industry*,
International Series in Operations Research & Management Science 310,
https://doi.org/10.1007/978-3-030-74154-9

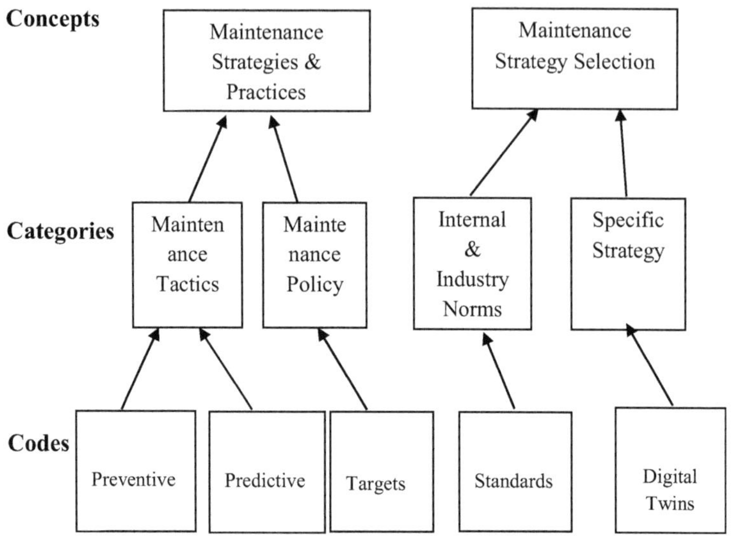

Fig. B.1 Categorization and linking of codes, categories, and concepts

In Fig. B.1, the codes represent the keywords identified related to the organization's asset maintenance strategy and practices. The categorization groups these keywords into different categories such as maintenance policy and maintenance tactics. Concepts represent the critical asset maintenance processes enable for asset maintenance management. Such detailed within-case analysis provides evidence of maintenance processes being adopted in the organization and their commitment to a comprehensive process to achieve maintenance excellence.

Therefore, identifying the underlying relationships between codes, categories, and concepts using the causal relationships ("is part of," "is associated with") is essential. Accomplishments mentioned by interviewees leading to the maintenance strategy processes are identified by creating appropriate Atlas.ti software codes. Figure B.2 depicts how causal relationships are identified from the interpretation of interview statements.

As shown in Fig. B.2, "standby equipment strategy" is the specific maintenance strategy adopted in critical processes that leads to plant availability of 98% on a 24 × 7 basis to achieve the operations-intensive organization's asset maintenance objective/policy. Similarly, the "previous experience" of asset maintenance managers is the cause of "preventive maintenance plans" in the organization. It implies that preventive maintenance planning is also carried out based on the organization's previous experience as maintenance engineers/managers. Accordingly, these codes are linked through Atlas.ti software by "leads to" and "is the cause of" and "is part of" relations as shown above.

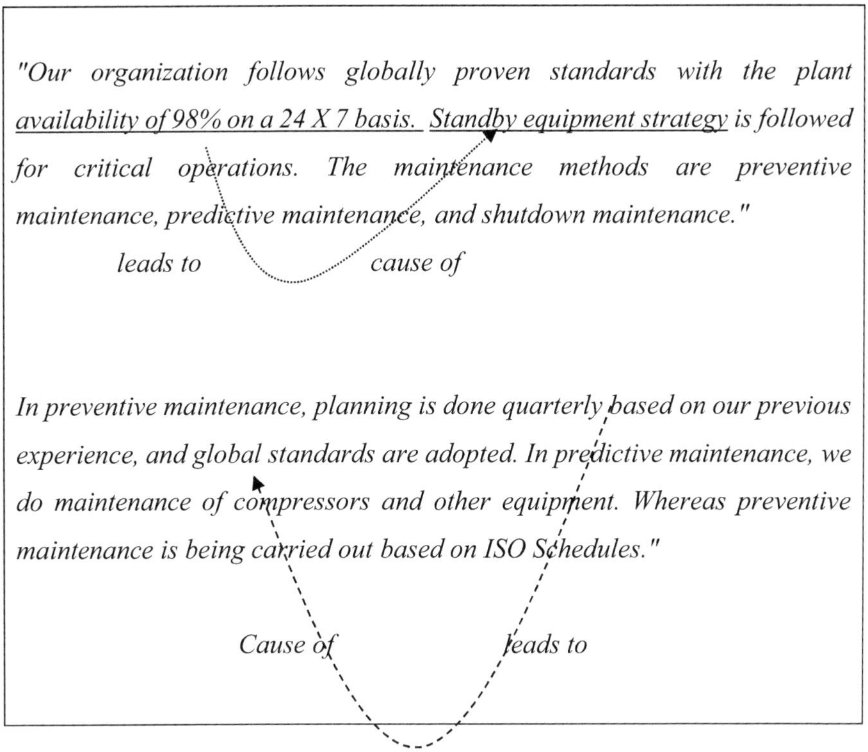

Fig. B.2 Casual relationships identified from interview statements

The abstraction of the codes (keywords) into categories and categories into concepts leads to the inheritance of relationships between codes. They are identified from the categories and codes. Continuing with the above example, the code "plant availability of 98%" is associated with "maintenance policy," which in turn is part of the "maintenance strategies and maintenance practices" concept. Similarly, the standby "equipment strategy" code is associated with the category "specific maintenance strategies for critical processes," which is part of the concept "selection of specific maintenance strategy for a particular operation process/equipment." Casual relationships such as "leads to" and "cause of" between the codes are categorized into categories and codes. The same is shown in Fig. B.3.

Data analysis is to be carried out based on the questions framed in the Baseline Survey Questionnaire (BSQ) in three phases, as depicted in Fig. B.4. The key objective of phase 1 is to understand the asset maintenance processes and contributing factors to asset maintenance strategy development and implementation using the above coding process through qualitative software. The phase 2 of data analysis concentrates on critical processes related to asset maintenance strategy selection

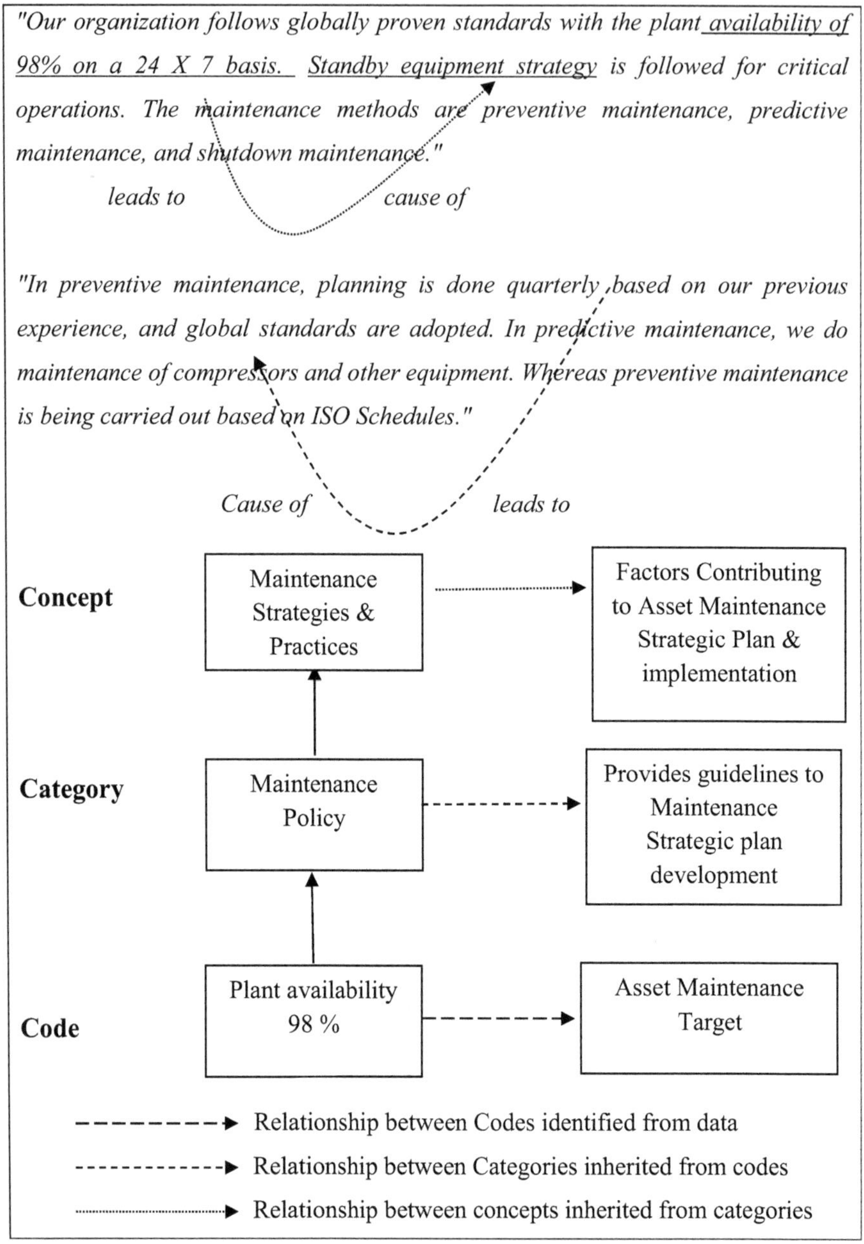

Fig. B.3 Relationships between concepts/categories inherited from codes (keywords)

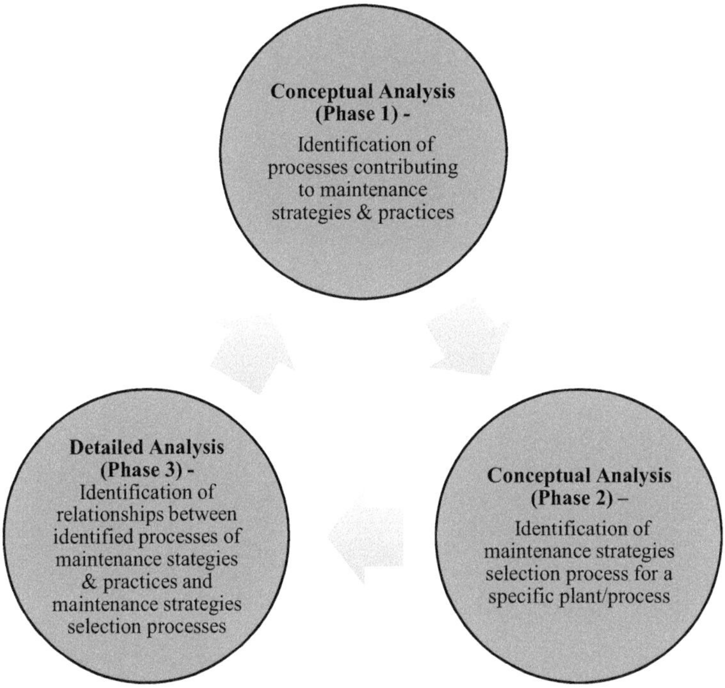

Fig. B.4 Phases of data analysis

regarding specific plant or equipment or processes as the case may be. The concepts and categorization of phase 1 are the inputs to this phase. Further, all identified processes must be evaluated and validated with asset maintenance practitioners during this stage to vet the data analysis outcomes. The complete systematic analysis resulted in a detailed description of existing asset maintenance practices in the organization. The above method helps the maintenance team develop a suitable asset maintenance strategic plan or revisit the existing asset maintenance strategic plan to incorporate suitable or necessary improvements to achieve asset maintenance excellence.

Appendix C

Assessment of Asset Maintenance Strategy and Practices of a Petrochemical Plant Using a Four-Stage Asset Maintenance Plan Development and Implementation (Case Study)

Petrochemicals are the derivatives of crude oil and natural gas. Olefins (ethylene, propylene, and butadiene) and aromatics (benzene, toluene, and xylenes) are the primary building blocks from which most chemicals and petrochemicals are produced. They are used in dyes, synthetic fibers, rubbers, plastics, pharmaceutical bulk drugs, industrial appliances, packaging industry, and detergents (surfactants).

The petrochemical production process consists of primarily two stages. In the first stage, naphtha, produced by refining crude oil or natural gas, is used as a feedstock and is cracked. Cracking (breaking of a long chain of hydrocarbon molecules) produces olefins and aromatics. In stage two, these building blocks are polymerized (made to undergo chemical processes) to produce downstream petrochemical products (polymers, polyesters, fiber intermediaries, and other industrial chemicals). The upstream integrated naphtha/gas cracker complexes are technology-intensive and enjoy the economy of scales. However, the downstream plastic processing industry is relatively fragmented globally and operates at lower than optimum capacity.

The industry suffers from high capital and energy costs, the shortage of natural gas, the lack of skilled workforce, low focus on value-added exports of end products, the cyclical nature of the business, zero import duty differentials between polymers, and feedstocks. This industry's key demand drivers are GDP growth, disposable income, country aspirations, and globalization. These megatrends get translated to increase demand for healthcare, packaging, white goods, automobiles, agriculture produce, and retail, and the margins are cyclical and typically follow a 6–8 years of period of troughs and peaks.

R. S. Velmurugan, T. Dhingra, *Asset Maintenance Management in Industry*, International Series in Operations Research & Management Science 310, https://doi.org/10.1007/978-3-030-74154-9

A petrochemical plant with a capacity of 400 KTA per annum is considered for this case study for assessing the asset maintenance strategic plan and its implementation in the organization. The petrochemical complex is producing polymer from natural gas. From natural gas, ethane (C2) is extracted through the gas processing plant, and then, C2 is cracked in gas cracker unit furnaces to convert ethane to ethylene. Then, ethylene is transported to downstream plants to convert to polyethylene grade high-density polyethylene (HDPE), low-linear-density polyethylene (LLDPE), and polypropylene (PP). The four-stage assessment process suggested in Chap. 6 is applied in this case study to understand asset maintenance management in the organization. The stage I of this process framework is applied to understand its current level of asset maintenance strategic implementation. A detailed study of the stage I process is presented below.

Semi-structured interviews were conducted using the Baseline Survey Questionnaire (refer to Appendix A) from the asset maintenance personnel working in the petrochemical plant, the organization's permanent employees. Data was collected from various sources such as on-the-job observation from asset maintenance personnel; semi-structured interviews; the organization's website; secondary data from the organizations; and informal discussions with asset maintenance team leaders, maintenance managers, maintenance engineers, and maintenance technicians. The total time taken for this stage is 2 months.

Based on the data collected, within the case, data analysis was undertaken to keep in mind the following research questions: how does the petrochemical plant plan and execute its maintenance strategies and practices to ensure smooth operation, and why is the specific maintenance strategy selected for a particular operation process/equipment? This analysis identifies various elements that form the basis for maintenance strategy selection and maintenance practices in the organization's petrochemical plant. Further, maintenance strategy selection processes being practiced in a petrochemical plant and specific maintenance strategies for a particular operation process/equipment were identified. The relationship between maintenance strategy selection processes and the identified elements for maintenance strategies and maintenance practices in petrochemical was analyzed. Detailed data analysis was carried out as per procedure given in Appendix B. Data analysis could be completed within a month.

The case's conceptual analysis was undertaken to identify various activities carried out by the maintenance managers for maintenance strategy selection and their practices in the petrochemical plant. These activities were then combined into categories/concepts based on maintenance strategies and maintenance practices. Interview data have also been analyzed in terms of the selection of formulated maintenance strategies.

A Qualitative Associative Network (QAN) is developed to understand the existing processes available in the organization under study in respect of asset maintenance strategy development and its implementation. The same is presented in Fig. C.1. The asset maintenance strategies and maintenance practices planning and execution in the petrochemical plant majorly depend on the various asset

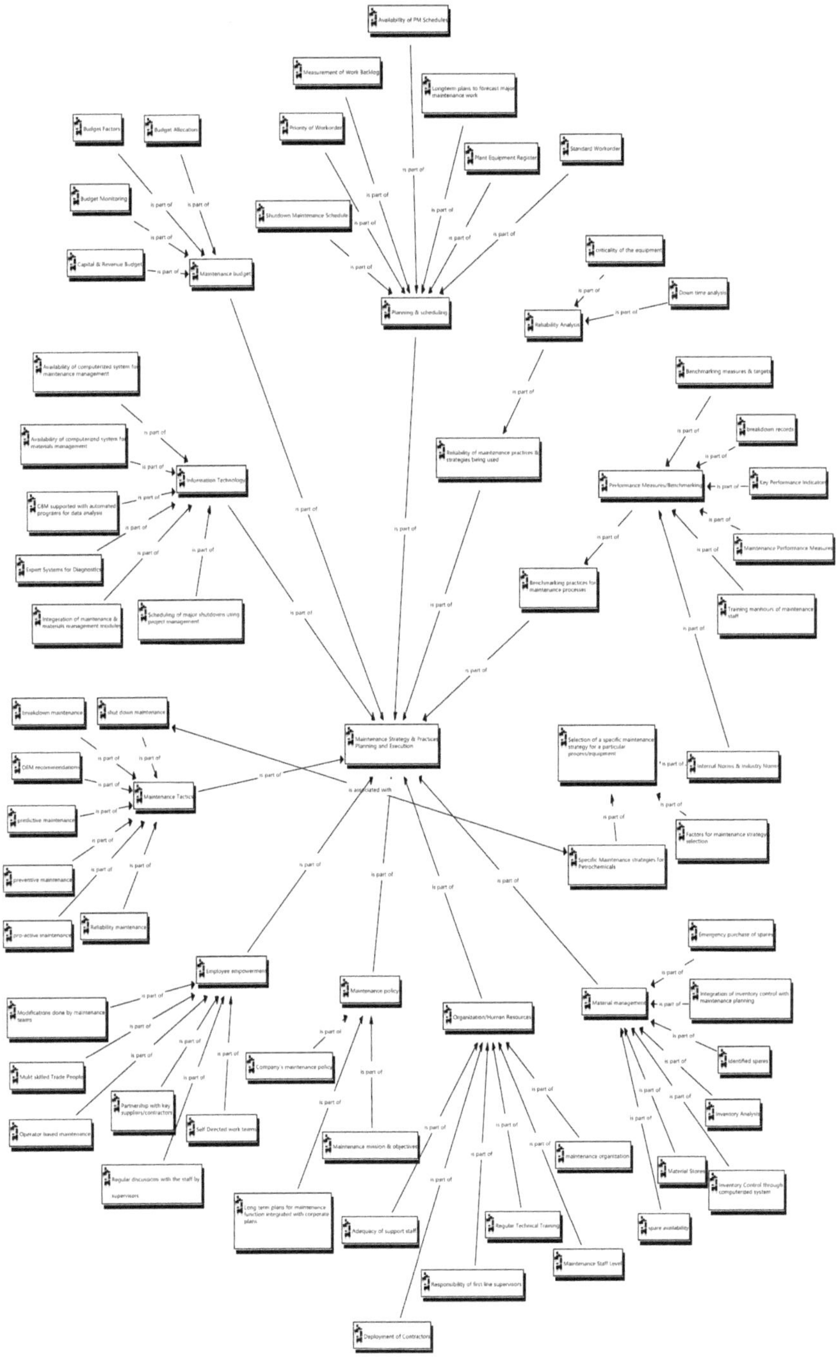

Fig. C.1 Qualitative associate network for petrochemical plant maintenance factors discussed in Chap. 2, contributing to maintenance strategies and maintenance practices in the organization.

The key findings of this case study are summarized as follows:

Predictive maintenance (Condition-Based Maintenance) and preventive maintenance (Time-Based Maintenance) are used in petrochemical plant-based equipment classification such as critical, semi-critical and non-critical. The Total Productive Maintenance (TPM) technique is not being practiced in the plant. Planned shutdown maintenance is the primary maintenance tactic used for the equipment/systems where standby is not available in the petrochemical plant. Proactive maintenance is the new maintenance tactic that emerged out from this case study. The same is to be further explored and implemented in the plant to assess the impact on maintenance excellence.

Reliability-Centered Maintenance (RCM) technique is not being practiced in the petrochemical plant. Therefore, the RCM metrics such as MTBF and value risk study are not used in asset maintenance in the organization. Benchmarking metrics are not available in the plant. However, the benchmarking exercise is being carried out by a consultant, and the results are expected in 3 months. The internal targets set by the corporate strategy department are treated as benchmarks for the petrochemical plant. The documentary evidence found that maintenance cost is benchmarked, i.e., 0.83% of gross block (inflation-adjusted).

Asset maintenance managers refer to corporate targets/operational targets as Key Performance Indicators (KPIs) for maintenance. There are no specific KPIs identified for measuring maintenance performance in petrochemicals. Maintenance performance is assessed through internal audits and external audits since these audits are being conducted based on industry standards followed in the country. The audit recommendations and their implementations are guiding the maintenance performance in the petrochemical plant.

Further, it is found that there is no evidence of awareness or usage of metrics related to labor cost/material cost from the interview data. However, maintenance cost is targeted by the organization as given above. There is no evidence of the following "average inventory turnover" concept in the organization. It is also found from the company's secondary data that the apprentice program is available as per statutory requirements. Maintenance challenges being faced by maintenance managers in their day-to-day jobs are grouped into technical issues, contract issues, stakeholder issues, internal issues, and unexpected issues. These factors play a crucial role in developing maintenance strategies and maintenance practices for the petrochemical plant. Presently, these challenges are managed by individuals/teams as and when required. The same has to be formalized and documented for better maintenance practice.

It is to understand that the critical maintenance challenges of petrochemical plant are time extension of scheduled preventive maintenance, time-based maintenance on an informed decision, obsolete technology of equipment and systems, earlier prediction of equipment failure, surprises during overhauling of equipment, deployment of contract workforce and workmanship of contract employees, the award of contracts, legal and statutory requirements, customer/supplier plant issues, quality of maintenance and inexperience maintenance personnel, and dependency on Original Equipment Manufacturer (OEM) support. These challenges are to be considered before formulating the asset maintenance strategic plan and while implementation of maintenance practices.

Finally, the selection of maintenance strategies for a particular equipment/operation process in a petrochemical plant includes factors such as the experience of maintenance managers and OEM recommendations. The petrochemical plant's specific maintenance strategies are shutdown maintenance, proactive maintenance, industry norms-based international and national level standards, and internal norms based on corporate-assigned targets and other practices such as the SAP system to integrate maintenance function.

The organization has introduced recent technological developments such as the Industrial Internet of Things in selected equipment, i.e., gas compressors and turbines. The up-gradation of control systems of this equipment and replacement of convention field sensors with smart sensors are in progress. The job is taken up in consultation with the Original Equipment Manufacturer (OEM) of the gas compressors and turbines.

The findings of the case study illustrate that interviewees considered the following asset maintenance factors such as performance measures, planning and scheduling, materials management, human resources, Information Technology, employee empowerment, maintenance tactics, reliability analysis, maintenance policies/maintenance budget, and maintenance challenges as essential in deciding on maintenance strategy selection process and maintenance practices in petrochemical plant in the organizations.

Further, interviewees indicated that the maintenance strategy selection process is not based on mathematical modeling using quantitative techniques and tools. It is based on previous experience of maintenance managers, the criticality of the system/equipment, OEM recommendations, maintenance policy guidelines, equipment complexity, equipment availability, process severity/complexity, and internal and industry norms.

MIX
Papier aus verantwortungsvollen Quellen
Paper from responsible sources
FSC® C105338

If you have any concerns about our products,
you can contact us on
ProductSafety@springernature.com

In case Publisher is established outside the EU,
the EU authorized representative is:
Springer Nature Customer Service Center GmbH
Europaplatz 3, 69115 Heidelberg, Germany

Printed by Libri Plureos GmbH
in Hamburg, Germany